# Applications of Biotechnology *in* Agriculture

# Applications of Biotechnology in Agriculture

P.K. Gaur

Oxford Book Company
Jaipur ▪ New Delhi

*Published By*

Oxford Book Company

**Regd. Off.** "Bony Residency", Gate No. 2
Opp. Tilak Public School, Vishweswariya Nagar,
Gopalpura Road, Jaipur - 302018, India
Ph: 0141 - 2761280, Telefax: 2761381
email - oxfordbookjpr@gmail.com
website - oxfordbookcompany.com

**Delhi Office:** 102, 1st Floor, Satyam House,
4327/3, Ansari Road, Darya Ganj,
New Delhi - 110002, India
Ph: 011-45652440

Published in India, 2025



ISBN : 9789350300336

Price : ₹ 2495/-



Typeset by Shivangi Computers, Jaipur

Printed and bound in India by Thomson Press Ltd.,
New Delhi

# Preface

Biotechnology is a field of applied biology that involves the use of living things in engineering, technology, medicine, and other useful applications. Modern use of the term includes genetic engineering as well as cell- and tissue-culture technologies. The concept encompasses a wide range of procedures for modifying living organisms just as to human purposes—going back to domestication of animals, cultivation of plants, and improvements to these through breeding programmes that employ artificial selection and hybridization. By comparison to biotechnology, bioengineering is generally thought of as a related field with its emphasis more on higher systems approaches for interfacing with and utilizing living things.

Since time began we humans have always relied on plants and animals to provide us with food, shelter, clothing and fuel, and for thousands of years farmers have strived to better these in order to continue providing us and to meet the needs of evolution. As the population grows and expands then so does the need for resources provided by plants and animals. With the population expected to reach over 10 billion by the year 2030 it is estimated that the world food production will have to double on the farmland that exists today, if it is to compete with the anticipated growth in the population. The ever-increasing needs of yields of crops and the decreasing crop inputs such as water and fertilizer can all be helped by biotechnology as can advancements in providing pest control that is better for our environment.

The discussion made in this book will help the students to understand and learn about the different aspects of agricultural biotechnology in a most comprehensive way to face the present and future challenges in this field.

*P.K. Gaur*

# Contents

## Chapter 1

# Technological Change in Agriculture

Technological change in agriculture has been the foundation of economic growth and development. By raising yields and continually increasing the number of people who can be fed by the output of one worker in agriculture, change in agricultural technology has permitted an increasing proportion of labour and other productive resources to be devoted to non-agricultural production. Plant and animal breeding and chemical and mechanical technology plus improved husbandry have caused continual structural change in farming. These technological stimuli have arisen as a series of overlapping waves to create a process of technological change, the momentum of which looks likely to be maintained by intensified adoption of information and biotechnologies.

As a result of the cumulative processes to date, farms have become larger and more specialized, and farmers are more highly trained and have substituted machines for animal and human power; farming in the 'West' has become one of the most capital-intensive of industries. At the same time agriculture has become more dependent upon industry, and even more industrialized-the intensive production of eggs, broiler chickens, horticultural products and pigs has many characteristics which belong to industrial production rather than to traditional farming. Goodman *et al.* talk of 'the industrial appropriation of the rural processes'.

Processes of marketing farm products have passed off the farm to be performed by sophisticated distribution and retailing sectors which increasingly dictate details of production to the farm sector, while at the other end of the chain an increasing proportion of inputs is supplied by the chemical, pharmaceutical and engineering industries. The process of 'industrializing' agriculture has been transferred to the less-developed countries (LDCs) and is epitomized by the so-called 'Green Revolution'. This has entailed expansion of irrigation systems using modern pumps, engineered dams and canals, plus the increased use of inorganic fertilizer, insecticides, herbicides, fungicides and power machinery; although at its heart has been the 'old' biotechnology of breeding new cereal varieties which give high yields when supplied with chemical inputs. Because much of the impetus for this 'revolution' has been western technology and farming know-how, particularly that of the USA, there are those who are critical of what they see as the LDCs' increasing technological dependence upon the West, its industries and research institutions. What is undeniably the case is that technological advance has been faster, and public and private expenditure on research in western agriculture greater than in LDCs taken as a whole.

Thus exportable surpluses from the West have increased with generous public support, while many LDCs have had to resort increasingly to food imports to meet their needs. While at this stage it is difficult to foresee exactly when, and on what commercial scale, the new agricultural biotechnology 'revolution' which is now brewing will have its effect, most commentators agree that it will intensify the industrialization of agriculture, and that it will increase the technological dependence of most LDCs upon western firms.

It may cause considerable disruption to the economies of some countries and will cause farming operations to diverge increasingly from the pattern associated with the country yeoman. Rural life in many more remote farming areas will be further threatened by the continued deterioration in the economics of ruminant livestock farming on upland and low-productivity pastures. It is these redistributive and structural effects that are the central focus here rather than any attempt to estimate the scale and speed of uptake of specific biotechnology inputs and processes. Although prospective biotechnological change in agriculture is not different in many fundamental respects from other processes of technological change, in that it will operate through the provision of new or modified inputs (seeds, hormones, etc.) and the creation of new markets in industry, it does raise new issues.

Are there dangers from releasing life forms which are engineered using new methods which are quite different from those associated with varieties developed by traditional plant-and animal-breeding methods? Even if the dangers can be objectively assessed and turn out to be minimal, will what may be perceived as man-made and therefore 'unnatural' products be acceptable to society as a whole?

## PRINCIPAL DIRECTIONS OF BIOTECHNOLOGICAL DEVELOPMENT

In a recent review of agricultural technology, the Office of Technology Assessment (OTA) of the US Congress defined biotechnology to include 'any technique that uses living organisms or processes to make or modify products, to improve plants or animals or to develop micro-organisms for specific uses-it focuses upon recombinant DNA and cell fusion technologies'. Longworth concurs with this definition, with the significant addition of tissue-culture techniques, an aspect of biotechnology which has great economic, and therefore social and political, potential impact. Despite debates about whether these or any other definitions are adequate, they do touch on the main aspects of biotechnology which deserve to be considered.

The aspect of this new biotechnology which most captures the imagination and stirs the greatest controversy is gene splicing or recombinant DNA techniques, which inspire researchers to consider the possibilities of producing reproducible animals and plants markedly different from current existing species, and referred to as transgenic species. Already in higher animals and plants recombinant DNA has produced transgenic forms which are being commercially exploited. Their agricultural significance is, however, so far limited, and commercial application is concentrated in highly profitable pharmaceutical and horticultural markets.

Genes have been introduced into several animal species which alter their protein synthesis to enable transgenic sheep to produce insulin in their milk, and rabbits to produce interferon. A completely different application has originated in Denmark for salmon, where it has proved possible to introduce germplasm which enables the salmon's physiology to handle heavy metals, which are normally toxic, so opening up new locations for farm fisheries.

Much current research is directed to conferring disease immunity on animals, and holds out the prospect of widespread commercial application. In plants one achievement has been the transfer of genetic resistance to antibiotics in the petunia, and another has been the introduction of storage-protein genes from French bean plants into tobacco plants. As yet commercial progress with recombinant-DNA technology in plants appears limited, but extensive opportunities beckon, particularly because plant research is less restricted by the ethical and animal-welfare concerns which apply to research on transgenic animals.

The most important commercial developments based on gene splicing have so far occurred with genetically much simpler microorganisms, and it is with these that the greatest short-to mediumrun commercial potential lies. Already genetically engineered micro-organisms are producing a variety of hormones, vaccines, enzymes and other proteins. Important examples are the production of insulin, vaccines for neo-natal diarrhoea in calves and piglets, and the bovine growth hormone BST identical to that produced in cows which can stimulate a 20 per cent increase in milk yield. BST is already a source of problems for legislators in the European Community, and has provoked strong reactions from the media and milk consumers. As far as larger animals, and cattle in particular, are concerned, it is developments in embryo transfer and in many processes for manipulating reproduction which hold out the prospect of continuing increases in yields, feed conversion efficiency and general economic efficiency. Already, apparently over 1 per cent of dairy calves in the USA are from embryo transplants, despite the high costs still associated with this procedure.

Widespread adoption of these sophisticated technologies would further distance livestock farming from its traditional rural simplicity, and from the natural mating of animals to produce offspring. It would place technical demands upon the operators which favour large-scale company farms capable of supporting a range of highly trained specialists. Longworth identifies new techniques of tissue and cell culture as having 'the potential for enormous advances in crop improvement in the next couple of decades'.

Those techniques 'can both increase the genetic diversity and greatly increase selection efficiency', and they permit innumerable plants to be reproduced asexually from single cells or small pieces of tissue. The particular technique known as 'callus culture' has been used for many years to clone highly valued horticultural plants such as orchids, and cloning of cuttings is widely practised for tree crops such as tea and palm oil as well as by millions of gardeners for garden plants. According to Longworth 'cell culture' of single cells has unexpectedly, and so far inexplicably, resulted in plants with different properties being regenerated from the same clump of parent tissue. Sugar cane, maize and potato plants regenerated in this way have been found which are resistant to important pathogens. This application of cell culture with the capacity to generate vast numbers of seedlings rapidly has the potential to simplify greatly the hitherto labourious procedures of plant breeding and selection.

However, in terms of current and immediate commercial importance, it is through micro-organisms that biotechnology has its greatest impact. Microbial fermentation processes have been of great commercial significance for centuries, for example in bread, wine and cheese production, and there is the prospect of considerable development. Two recent examples of new processes indicate the sorts of impact that such developments can have. In the late 1960s genetically engineered bacteria were developed which were able to digest corn starch to produce high-fructose corn syrup (HFCS) and which left as a residue corn gluten, which is now an important protein feed for livestock. HFCS has made considerable inroads in the United States and in Japan.

It has largely replaced sugar as a sweetener in Coca Cola as well as many other food and drink products. This has helped depress sugar prices to Third World growers. In the European Community steps have been taken to prevent HFCS and other new powerful sweeteners from being produced and from undermining the market for domestically grown beet sugar, which is supported by the Common Agricultural Policy. A second important application of microbial fermentation has been the production of ethanol from sugar cane in Brazil and from maize in the USA.

The commercial viability of these processes is critically dependent upon the price of oil as the main non-renewable source of fuel, and, to date, massive subsidies have been required to maintain the Brazilian and USA ethanol programmes. Longworth states, however, that a new biotechnology, Sucrotech, is being patented, which will not only reduce the cost of producing ethanol from sugar cane, but will simultaneously produce fructose at a cost which will be competitive with HFCS and may thereby reclaim part of the sweetener market for sugar cane.

That such possibilities are in prospect indicates how volatile the future might be; biotechnology has tilted the competitive balance from sugar cane to maize, causing economic pressure and even disruption to sugar-cane-dependent economies and may in future switch it back again. In the long run the capacity to ferment fuels microbially from renewable agricultural feedstocks points to an important long-term reorientation of agriculture if non-renewable oil becomes uncompetitively expensive as a fuel for cars and feedstock for certain chemicals.

It suggests that eventually there will be an increased emphasis on agricultural production of industrial feedstocks at the same time as continually increasing food output will be required to feed the expanding world population. All this will require considerable increases in agricultural productivity, to which biotechnology will increasingly contribute, but it will at the same time impose great strains on the natural environment and the structure of agriculture.

## PRIVATE-SECTOR CONTROL OF BIOTECHNOLOGY DEVELOPMENT

It has been the role of the public sector to undertake research and development of agricultural technology as a public service to firms which might profit from translating that R & D into a commercial product or process, to farmers who might profit from adoption of the technology, and perhaps most importantly to consumers at home and abroad who benefited from the lower prices resulting from greater abundance. This was particularly true of the phase of change dominated by improvements in plant and animal breeding, the 'old' biotechnology. While the public sector had a role in basic research for chemical and mechanical technologies, the benefits of research expenditure in these areas were easier to capture by private companies investing in R & D, so that increasingly the public sector has taken a smaller role in these areas although maintaining a strong regulatory role in regard to agricultural chemicals in particular.

Where it is impossible to prevent others from escaping payment for the research costs, either because the product is easily copied (seeds which can be regenerated by farmers or other firms) or because proposals for new methods can be readily implemented, private firms are understandably unwilling to invest. Machines, insecticides, fungicides and other manufactured inputs do lend themselves more readily to private exploitation. Nevertheless the returns to R & D in these products do depend upon the degree of difficulty potential competitors would have in copying the product. In some cases there are inherent technical difficulties in copying the process, or it would be prohibitively expensive, but in other cases it is the ability to obtain patent protection which creates legal barriers to potential competitors' ability to become 'free-riders' and which protects incentives to private investment in R & D.

Traditionally, however, it has been impossible for plant and animal breeders to obtain patent rights for their products, which is one reason why public-sector R & D has remained so important in this area. For, as under the European Patent Convention (EPC) of 1973, it has been judged impossible for plant varieties to satisfy one of the key criteria to qualify for a patent, namely proof of 'an inventive step'. The application of standard breeding practices to generate new varieties by crossing existing plants or animal strains has not been deemed to be invention.

Thus under the EPC one set of exclusions from patentability is 'Plant or animal varieties or essentially biological processes for microbiological processes or the products thereof.' In the absence of patent rights plant-breeding firms in particular have worked hard to obtain other means of protection and royalties for their products. The history of the development of Plant Breeders' Rights (PER) is presented by Mooney, and is of particular interest because of the concerns he expresses about the consequences of allowing the basic genetic stock, which underpins agriculture and hence our whole society, from becoming private rather than public property. While Mooney's concerns are expressed in relation to 'old' biotechnology varieties of plants, they are of particular importance with respect to the products of the 'new' biotechnology.

For bio-engineered products are capable of meeting the inventive-step criterion of patentability, and both micro-organisms and transgenic animals have now been patented in the USA. Before examining developments in the seed industry it is worth touching upon the implications of changes in research policy which stress increasing reliance upon private R & D to develop and exploit the new biotechnology and other technologies.

In the UK government has decided that it should withdraw from funding what it terms 'near-market research', that is, research beyond the basic phase and which is preparation for commercial exploitation. This policy is based in part upon the arguments that industry should be investing more heavily in R & D, and that research strategies at the near-market stage should be driven by assessments of likely commercial success which can best be made by the firms involved, and will therefore lead to greater efficiency in allocating research funds. This has led to the closure of a number of government-financed agricultural research institutes, the scaling down of others and the sale of the National Seeds Organization by auction to Unilever; Unilever won against competition with BP and other major public companies. This deliberate attempt to switch an increasing proportion of agricultural R & D expenditure from the public to the private sector carries with it a number of risks.

In the first place there is a controversy about whether there is under-investment in R & D so that the returns to extra expenditure are high, or whether the converse is the case. If there is under-investment, then withdrawal of public support for applied research will exacerbate this, as the private sector will only undertake R & D expenditure on those products and processes from which exclusive benefits can be captured by the investor.

Thus in relation to biotechnology in the USA, where again public support for applied biotechnology research is relatively small, Stallman and Schmid (1987) argue that there will be emphasis on technologies which are applied in factory conditions where secrecy and control can be maintained. For technologies which cannot be confined to factories, such as seeds, Stallman and Schmid state that 'Firms are also considering mechanisms which "scramble" the genome of a plant in the second generation', in order to prevent farmers from reproducing seed with the enhanced, engineered characteristics. Clearly such actions are designed to frustrate the maximum spread of benefits from the technology and to maximize private profit for companies investing in research. Another facet of commercially orientated biotechnology research is that it will aim at the most important crops and livestock products, those produced by the largest farming units, and those which are most heavily subsidized. In the latter case this will worsen the budgetary problems of adjusting agricultural policy in OECD countries. The probable neglect of minor crops, difficult habitats and small farmers means that the public sector will have a defined role in agricultural technology research, but one in which it will be relegated to the second division and where it is unlikely to prove successful in terms of the commercial yardstick of rates of return which is increasingly emphasized by public-research policy.

## GENETICALLY ENGINEERED PLANTS AND FOODS

At a symposium at the University of Washington Law School in October 1993 (on the eve of President Clinton's Asian Pacific Economic Council 'Summit'), when discussing the 'Future of Intellectual Property Protection for Biotechnology in the US, EC, and Japan', Joseph Straus of the Max Planck Institute warned the participants against 'ethics and other irrational considerations'. Indeed, except for a very narrow band of 'moral dilemma' situations which are the stock-in-trade of professional biomedical ethicists, the ethical aspects of genetic engineering have been routinely ignored by government policy-makers and corporate technology promoters.

My thesis is that it has been the corporate promoters and their governmental handmaidens who have been 'irrational' in their systematic refusal to acknowledge the environmental and ethical considerations of genetic manipulation. Technologies, by definition, are neither acts of God nor nature; they are the embodiment of specific human purposes and intentionality. Far from being inevitable, they are researched and developed by those entities with sufficient power to mobilize social institutions to bring the technology into being, according to their own goals and normative considerations. 'Biotechnology is not neutral.

It shares the propensity of modern materialistic science to desacralise, dominate and manipulate life. It reduces all living things to a mechanism which it can manipulate according to engineering standards.' Genetic engineering is a technological process for undertaking activities which do not, and cannot, occur in nature. In this sense it is perfectly legitimate, therefore, to label genetic engineering as 'unnatural'. The European Community has incorporated this concept in its definition of genetically modified organisms, and it has been adopted by the UN Environment Programme (in the working of its Fourth Expert Panel set up under the Biodiversity Convention as part of the follow through to the United Nations Conference on the Environment and Development held in Rio de Janeiro in June 1992): 'organisms in which the genetic material has been altered in a way that does not occur naturally by mating and/or natural recombination'. Thus, hybrids and other modified organisms obtained by traditional breeding techniques are excluded from the definition of genetically modified organisms.

It is important to note that the process as well as the products are novel. The promoters of the new technology believe that a concern for the process itself is specious, downplaying its novelty whenever confronted by discussions of regulatory oversight. However, one of the bases of the decision of the US Supreme Court in the Chakrabarti case (which, 5 to 4, upheld the patentability of genetically engineered microbes, and which has been unquestionably accepted by foreign governments as dispositive of the issue of patentability of genetically engineered life forms) was that the element of 'novelty' required by the patent law was to be found in the process by which the microbe had been created.

One reason why we cannot ignore the powerful novel aspects of the processes of genetic manipulation is because we are not omniscient as to what will, in fact, happen when we alter genomes. Genetic manipulation is not like a child's game of Legos or Tinkertoy, in which parts can be rearranged or linked up with only simple mechanical and relatively predictable consequences. In calculating any risk from a transgenic organism, one should consider four elements: the host organism, the foreign genes, the interaction between the foreign genes and the rest of the genome, and the environment in which the organisms will be used.... [in regard to the last two elements] the literature contains many examples of genetic manipulations where inserted genes did not respond in their new environments the way they did in their old ones or where alterations with one part of the genome caused surprising activity in other parts of the genome.

There is an element of arrogance to scientific assertions which assure us that the process itself poses no risks, since scientists know so little about actual ecosystems; for example, I have been told by agronomists that over 80% of the organisms which can be identified in a soil sample from my garden are completely unknown to the scientific literature. Indeed, the Ecological Society of America itself has specifically warned about the problems presented in our lack of ecological knowledge and the resulting consequences from releasing (intentionally or accidentally) genetically altered organisms into ecological systems.

## FOODS

In May 1992, the US Food and Drug Administration, responding to corporate pressures to remove the prospect of regulation of genetically altered foods, issued a set of rules which have largely left responsibility for protecting the public health and safety in the hands of the industry, permitting products to be marketed without scrutiny unless the industry indicated to the agency that it believed governmental oversight was justifiable.

However, the proposal as published by the FDA in the Federal Register perversely noted several important problem areas: creating allergens, additions of genes from sources which might violate religious and cultural norms (of vegetarians, Jews or Muslims, etc.), and implications for animal welfare and well-being (for example, the incorporation of human growth hormone-producing gene into a pig's genome produced a highly arthritic animal).

Indeed, the commissioner of the FDA, David A. Kessler, and his colleagues noted the possibility that genetically engineered food might 'contain high levels of unexpected, acutely toxic substances'. In order to begin to address ethical issues presented by the genetic modification of foodstuffs, we need to have some ideas of the purposes for which these modifications are performed. (For example, if we wish to approach an analysis on utilitarian cost-benefit grounds.) Despite a great number of general statements, human nutrition and hunger do not appear to be the actual driving forces behind the development of genetically engineered foods.

Other than claiming increased shelf life (which can be indirectly related to nutrition in the sense of reducing spoilage or the consumption of foods that have begun to spoil), there is hardly any indication at all that genetic engineering is being directed to create nutritious substances out of nonnutritious ones; similarly, there are few real instances of genetic engineering reducing the cost of foodstuffs, or increasing the quantities of foods actually available to populations that are hungry because of the non-existence of consumables (as opposed to hungry because they lack money to buy sufficient foods, no matter how produced). Indeed, early genetic manipulations seemed perversely designed to minimize the achievement of such goals: the creation of 'herbicide tolerant plants' which do not reduce the applications of dangerous chemicals to agricultural fields (which might occur if genetic manipulations of food crops were designed to make them resistant to insects, fungi, or disease) but instead permit higher levels of chemical application, or the introduction of recombinant bovine growth hormone to produce more milk at a time when developed countries suffer from milk gluts and have instituted programmes to kill cows and physically dump milk (milk as a commodity is price inelastic; quantity increases do not lead to price decreases).

The goals of genetically engineering foods seem to be more closely tied to increasing the economic gain and power of the corporations involved in food production (for example, the development of herbicide-tolerant farm crops has been led by corporations which manufacture herbicides), making production easier for corporations, reducing or altering packaging and transportation costs for the corporations, etc.

**Environment**

Industrial societies around the world are faced by massive pollution legacies from the previous emphases on chemical and nuclear industrial activities. The inability of the US Department of Energy to find a suitable site for a longterm nuclear waste repository illustrates our very poor track record for dealing with the scientific, technical, economic, and socio-political aspects of prior technological 'revolutions'.

The environmental problems posed by genetically engineered organisms are likely to be substantially more intractable than those posed by these earlier instances of pollution because genetic wastes multiply, migrate, and mutate. A genetically engineered organism once free in the environment is impossible to recall. (Impacts which are irreversible must be considered with higher scrutiny than those which can be undone.)

Examples of environmental problems which need to be assessed include the risks of transgenic crops themselves becoming weeds; the risk of gene flow to wild relatives which might become weeds or pests; the growth of antibiotic resistance in species (particularly animals), since resistance genes are used as markers for the genetic engineering sites; the problem of exotic or non-native species taking over ecological niches (such as gypsy moth, starling, kudzu, rabbits in Australia, etc.) leading to extraordinary economic losses as well as ecological ones; the restriction (rather than increase) of biodiversity by selective advantageous breeding of transgenic organisms as opposed to natural ones, or the predatory results of expanding exotic species (such as the Dutch Elm micro-organism destroying the American Elm and severely restricting the biodiversity of certain areas in the Northeast United States); health issues (both of plant and animal species, as well as humans - worker health and safety as well as community security); and the occurrence of the completely unexpected, the inability to eliminate uncertainty (for example, the late 1993 floods in the Mississippi valley included the flooding of a field of genetically engineered corn and the dispersal of this plant material to unknown sites within the thousands of square miles of downstream flood plain). Prospective ecological assessment is not being performed, despite the fact that its need has long since been recognized. And economic priorities, particularly those accruing directly to the promoters, as well as the nebulous spur of 'competition' with foreign countries, oftentimes overwhelms any interest in doing environmental impact analysis.

## DEVELOPMENTS IN THE SEED INDUSTRY

Many major food crops upon which we depend originated and were first cultivated in Third World countries rather than in the industrialized countries where they are most productively exploited today and in which they have been improved by traditional plant breeding. The potato originated in South America, wheat in Ethiopia and the Near East, important maize varieties in South America, and rice in Asia. Moreover the so-called Vavilov centres, which contain the greatest variety of living species and which have been the source of much important gene material, are mainly in Third World countries. The wealth of the industrial countries owes much to their past ability to exploit agriculturally crops and animals originating in other countries.

In the colonial era the success of British scientists in smuggling rubber plants from Brazil to Sri Lanka, Singapore and Malaysia and the comparable more complex route by which coffee was introduced to Latin America from Ethiopia are instances of colonial powers obtaining significant wealth by exploiting plant material from the Third World. In one way or another the process by which institutions and firms in developed countries have continued to collect new plant and animal varieties from less-developed countries has continued. But what is still a major source of friction is the procedures by which legislative protection is developed to give companies in the richer countries commercial rights to exclusive exploitation of varieties developed from plants freely collected from other countries, and to extract high rates of profit from the sale of those varieties. Most extreme is the situation whereby under the USA Plant Patent Act of 1930, which covered asexually propagated plants such as certain fruits, flowers, ornamental shrubs and trees, it has been possible for plants discovered and smuggled out of other countries to be given patent protection in the USA. The key to this is that the Act accepted the unfeasibility of requiring proof of an 'inventive-step' for the granting of a patent, and employed the criterion that as far as can be determined the plant variety is new. Since some novel variety found in the rain forests of South America may satisfy this criterion, and the source of the plant can be concealed from the authorities, the theft and smuggling of plants can be said to have been encouraged.

Seed companies in developed countries have successfully pressed their case for protection, and have progressively extended Plant Breeders' Rights (PBR) through both national legislation and the Union for the Protection of New Varieties of Plant (UPOV), which is an international agreement among signatory countries to honour a system of varietal rights. In the case of food crops, in the absence of patentability, PBR operates by registering distinct varieties. Pressure has been exerted upon less-developed countries to become signatories of UPOV and thus acknowledge the rights of breeders, most of whom are from developed countries.

However, the majority have stalled or rejected membership, and have been fundamentally unhappy about accepting the principle of creating monopoly rights for others over plant material which in some cases originated in their own country. The whole issue of intellectual property rights has been raised to a prominent political level by the USA's insistence that it be included as one of the thirteen areas for negotiation in the Uruguay Round of multilateral bargaining of the General Agreement on Trade and Tariffs. While the issue is a much broader one than that of plant and animal breeders' rights, it reflects a basic political conflict between the richer countries, determined to obtain secure returns on R & D expenditures by companies, and LDCs struggling to catch up and anxious to avoid having to pay royalties on products protected by PBR and patents.

This conflict has broader moral and ethical dimensions when it relates to the issue of charging royalties for the seeds of food crops to countries where malnutrition is widespread. Against the previous argument has to be balanced the need of those incurring R&D expenditure to capture enough of the returns to provide an incentive for R & D. At least that is certainly the case where R & D is to be undertaken by the private sector, which is politically the increasingly preferred solution in the UK, the USA and elsewhere. This, however, leads to another area of concern, which is that a small number of large multinational chemical and pharmaceutical firms have increasingly come to control the seed industry. Companies such as Shell, Sandoz, Dekallb-Pfizer, Ciba-Geigy and BP have become owners of most of what were small independent seed companies. This linking between chemicals and seeds within firms has a strong commercial logic. It facilitates the marketing of packages consisting of seed varieties plus the appropri-ate fertilizers, insecticides, etc.

It also allows the companies concerned to integrate their plant breeding, biotechnology and chemical research. One particular synergy for these companies will be to bioengineer varieties of major food crops to resist their own herbicides and weedkillers. This would permit chemical weed control to be extended to large-scale crops where this is not now possible, and is just one way in which the direction of biotechnolo-gical research may be influenced (biased) by the combination of forces giving a small number of companies extensive control over the basic means of agricultural production.

There are many other issues which could be touched on regarding corporate influence over the development of agricultural biotechnology. It is, however, understandable that the spokesmen for LDC countries, which are profoundly concerned about their dependence upon western technology and exercised by what is seen as colonial and post-colonial exploitation by the West, should be alarmed by the prospect of new rights being created over seeds and other biotechnological products.

Certainly there are widespread fears that they will only gain access to the fruits of biotechnology on unfavourable terms and that their relative dependence on and subservience to western industry will be increased so that benefits to them will be small. There are many western critics who would agree with this, and that the system by which agricultural biotechnology is delivered will favour the 'haves' rather than the 'have-nots'.

## THE FUTURE OF AGRICULTURAL BIOTECHNOLOGY

There are divergent views about the rate of uptake of new agricultural biotechnologies and their impact. Kalter and Tauer and the US Office of Technology Assessment anticipate comparatively rapid take-up, while others (Buckwell and Moxey and Farrington do not foresee much impact until the next century, that is until ten to fifteen years have elapsed. The caution of the latter commentators is probably justified, for there are various hurdles to be jumped before biotechnologies can gain acceptance. One hurdle to be overcome is the economic one. Self-evidently there must be some economic advantage to farmers or agro-industry from adopting the technology. Frequently what captures the scientific imagination turns out to be economically unviable in use, although it is only with use that efficiency can be improved and costs brought down. A more demanding test still for biotechnologies will be to obtain legislative approval and public acceptance. These are intimately connected.

In the case of machinery developments, provided safety standards are observed there are no legal impediments to developing new and improved machines, since there are no obvious problems of hazard to the public and therefore public acceptance. Chemical insecticide, fungicides and herbicides do, however, pose much greater problems because of concerns with toxicity, and elabourate testing and licensing procedures have evolved after a relatively haphazard set of procedures in the 1950s and 1960s, when widespread use was made of DDT and organo-phosphorous compounds without adequate recognition of their toxicity. In significant part, because of the problems which have been experienced with agro-chemicals, the licensing of biotechnologies will inevitably be based on testing at least as stringent as that which now exists for chemicals.

Indeed the licensing procedures are likely to be more stringent because of public and scientific concerns. Public acceptance of biotechnology in the food chain will be made more difficult by confusion about differences between biotechnologies. In the early 1980s there was a crescendo of concern about the use of steroids to promote faster liveweight gain in calves and cattle. The public outcry eventually led to the banning of such hormones for meat animals; but the legacy is a profound distrust of and hostility to new products such as synthetic Bovine Somatrophin (BST), which has the capacity to increase the milk yield of those dairy cows which are 'relatively' deficient in what is a naturally occurring hormone.

Although initial scientific results are favourable, and some minor doubts remain, the major influence leading the European Commission to impose a moratorium on the use of BST until the end of 1990 at least is concern about public perception; the Commission has stated 'It would be a serious setback to producers and to the Community's milk policy were present [positive] trends in consumption to be reversed as a result of adverse consumer reaction. With newly bioengineered plants and animals scientific and public concerns are emerging which are of a different order from those associated with previous biotechnology. One relates to the possibility that crop failures may become more frequent if biotechnology leads to a reduction in genetic diversity in crops being grown; such a tendency has already been observed in relation to the Green Revolution technology for wheat. There are also concerns about upsetting the balance of nature in unforseen ways, akin to the unanticipated consequence of introducing rabbits into Australia and then trying to control these by introducing myxomatosis.

Thus questions arise such as what happens if herbicide resistance introduced into a commercial crop plant transfers itself by cross pollination into a closely related weed species, or indeed if the herbicide-resistant crop should colonize wild habitats. There are, of course, more lurid and absurd notions bandied about in the popular press which are similar in nature to the attacks made on Darwinism.

The questions, both absurd and real, raised in relation to agricultural biotechnology will undoubtedly slow the rate at which it is adopted. Nevertheless instances of adoption are increasing: BST in the USA, a bioengineered baker's yeast in the UK, bioengineered sheep producing insulin, etc. As these become more widespread and increasingly affect lower-valued, bulk agricultural products, so the impacts of biotechnology on structural change will increase. The balance of economic power will switch increasingly to industry, to high-technology large farms and against smaller farmers in disadvantaged regions and countries. That, unfortunately, is a seemingly inevitable consequence of what we consider to be economic progress.

Chapter 2

# Genetically Modified Food Crops

## EVALUATION OF GENETICALLY MODIFIED FOOD CROPS

Genetic modification, otherwise referred to as recombinant DNA (rDNA) technology or gene-splicing, has proven to be a more precise, predictable and better understood method for the manipulation of genetic material than previously attained through conventional plant breeding. To date, agricultural applications of the technology have involved the insertion of genes for desirable agronomic traits (e.g. herbicide tolerance, insect resistance) into a variety of crop plants, and from a variety of biological sources.

Examples include soybeans modified with gene sequences from a Streptomyces species encoding enzymes that confer herbicide tolerance, and corn plants modified to express the insecticidal protein of an indigenous soil microorganism, Bacillus thuringiensis (Bt). A growing body of evidence suggests that the technology may be used to make enhancements to not only the agronomic properties, but the food, nutritional, industrial and medicinal attributes of genetically modified (GM) crops.

Regulatory supervision of rDNA technology and its products has been in place for a longer period of time in the United States than in most other parts of the world. The methods and approaches established to evaluate the safety of products developed using rDNA technology continue to evolve in response to the increasing availability of new scientific information.

As our understanding of the potential applications of the technology is broadened, the safety of products developed using rDNA technology and the potential effects of introduced gene sequences on human health or the environment will be more closely scrutinized. In fact, much of the knowledge acquired during the commercialization of the products of rDNA technology in agriculture is now finding application in evaluating the safety of products developed through more conventional means. The objective of this chapter is to provide the reader with an overview of the significant events leading up to the present, science-based, regulatory framework that exists for the safety evaluation of GM food crops within the United States. An attempt has been made to discuss concerns over the sufficiency of existing regulations, as well as to highlight recent initiatives taken by federal regulatory agencies to address them.

Through better communication of how the regulatory process functions within the United States, it is anticipated that current and future applications of rDNA technology in agriculture will be met with a greater level of understanding and acceptance.

## HISTORICAL PERSPECTIVE

rDNA technology was first developed in the 1970s. The initial response of the scientific community, including members of the National Academy of Science (NAS), to the prospects of rDNA technology, was to postpone any further research involving the technology until the potential risks to human health and the environment could be evaluated.

Researchers attending the International Conference on Recombinant DNA Molecules in 1975, otherwise known as the Asilomar Conference, tried to establish a scientific consensus on how best to self-regulate emerging applications of the technology. The conditions and restrictions that were proposed at this conference have formed the basis by which federal guidelines and policies for rDNA technology research were drafted within the United States.

### National Institutes of Health (NIH)

The National Institutes of Health (NIH) was the first federal regulatory agency to publish their interests in evaluating the safety of rDNA technology in 1976, in the form of guidelines for the conduct of research. Because of the uncertainties that existed at the time, all research into the potential applications of rDNA technology was limited to the confines of federally funded labouratories under NIH control. After continued research, and a more careful assessment and monitoring of the risks, a set of less restrictive guidelines was published in 1978.

However, the environmental release of organisms developed using rDNA technology outside the confines of controlled labouratory conditions was prohibited unless otherwise approved by the NIH director. In the early 1980s, the NIH established an rDNA Advisory Committee (RAC) to review all data and experience gained with applications of the technology under its control. Based on recommendations of the RAC, a more relaxed set of research guidelines was published by the NIH in 1983.

The NIH approved the first environmental release of an organism developed using rDNA technology (ice-minus strain of Pseudomonas) in 1983. In response, they were criticized for failing to prepare a statement or assessment of the environmental impact of their regulatory decision as required under the National Environmental Policy Act (NEPA). Once the legal controversy had subsided, all responsibility that the NIH had for regulating the environmental introduction of GM organisms was relinquished.

Nevertheless, NIH guidelines continue to be referenced in assessing the safety of rDNA research performed within industry, federal and other state labouratories. However, it was unclear which federal regulatory agencies would be responsible for ensuring the safety of the products developed using rDNA technology.

## OFFICE OF SCIENCE AND TECHNOLOGY POLICY (OSTP)

In response to a need for clarification, the Office of Science and Technology Policy (OSTP) began work on the development of a policy to establish a federal regulatory framework for evaluating the safety of products developed using rDNA technology. Following an opportunity for public comment, OSTP published a final version of the 'Coordinated Framework for Regulation of Biotechnology' (Coordinated Framework) in 1986.

The policy provided the basis by which federal regulatory agencies got involved in evaluating the safety of products at later stages of commercial development at that time. According to the Coordinated Framework, the products of rDNA technology should be regulated on the basis of the unique characteristics and features that they exhibit, not their method of production. The products of rDNA technology were considered to pose risks to human health and the environment similar to those posed by conventional products already regulated within the United States.

As a result, no new federal regulatory agencies or regulations were required. The Coordinated Framework did not, however, rule out the possibility of the development of new guidelines, procedures, criteria or even regulations to supplement or alter the scope of existing statutes for the products of rDNA technology. The Coordinated Framework identified three federal regulatory agencies within the United States: the US Food and Drug Administration (US FDA), the US Department of Agriculture (USDA) and the US Environmental Protection Agency

(US EPA), as having primary responsibilities for evaluating the products of rDNA technology under development at that time. In 1992, the OSTP released another document entitled, 'Exercise of Federal Oversight within the Scope of Statutory Authority: Planned Introductions of Biotechnology Products Into the Environment', outlining the proper basis by which federal regulatory agencies were expected to exercise their regulatory authority. As with conventional products, dependent upon the intended use and function, more than one federal regulatory agency may share an interest in evaluating the safety of a product developed using rDNA technology.

If more than one federal regulatory agency has an interest, lead agencies are identified as being responsible for coordinating activities to limit any potential duplication of efforts. Although federal regulatory agencies worked independent of one another, it was realised that close working relationships would need to be established in order to evaluate effectively the safety of products developed using rDNA technology.

Recently, the OSTP teamed up with the White House Council on Environmental Quality (CEQ) to perform a six-month inter-agency evaluation of the federal regulatory agency responsibilities in evaluating the environmental safety of products developed using rDNA technology. A case-study approach for a variety of different classes of products developed using rDNA technology was used to evaluate the level of federal regulatory agency involvement, to identify strengths, weaknesses and areas of potential improvement. The review concluded that none of the previously approved products of rDNA technology has had any significant negative impact on the environment.

Although all the case studies were published, OSTP/CEQ failed to reach a consensus on issues relating to the relevant strengths and weaknesses of the existing regulatory structure within the time allotted for the completion of its review. A review of the case studies published provides a comprehensive interpretation of the responsibilities of each federal regulatory agency in ensuring the safety of the products developed using rDNA technology.

## NATIONAL ACADEMY OF SCIENCES (NAS)

The National Academy of Sciences (NAS), and its operating arm, the National Research Council (NRC), have served as a primary source of scientific, technological, human health and environmental policy advice during the development of regulatory approaches for the safety evaluation of products developed using rDNA technology within the United States. In 1987, the NRC published a report concerning the potential human health and environmental hazards associated with the commercial introduction of GM organisms, entitled Introduction of Recombinant DNA-Engineered Organisms into the Environment: Key Issues. The risks associated with the introduction of GM organisms were considered to be essentially the same in kind as those associated with unmodified organisms.

In other words, rDNA technology did not appear to introduce any unique risks as compared to the products that had been developed using more conventional methods of genetic modification. In reaching these conclusions, the NRC performed an evaluation of the similarities and differences in the properties exhibited by products developed using a variety of different techniques. To this day, the conclusions of this report continue to be referenced by the regulators and developers of GM crop varieties worldwide. A subsequent NRC report, entitled Field Testing Genetically Modified Organisms: Framework for Decisions, also reached similar conclusions, but provided additional guidance as to how regulatory decisions concerning the introduction of GM organisms should be made. The NRC recommended that regulatory decisions concerning the introduction of GM organisms should be made on a case-by-case basis.

Consistent with the Coordinated Framework, the NRC did not consider the nature of the process used for the genetic modification of an organism to be a useful criterion for determining whether a product requires less or more regulatory oversight. As a result, no valid reason existed to regulate organisms genetically modified via modern techniques (e.g. rDNA technology) any differently from organisms genetically modified via more conventional means.

Similar conclusions have been published in the reports of international standards-setting organizations. In retrospect, within both reports, the NRC acknowledged that modern and conventional methods of genetic modification are not without risks to human health or the environment, and as a result, neither could be considered inherently more risky. As a result, regulatory decisions concerning the safety of the products of rDNA technology need only to take into consideration the specific characteristics exhibited by a particular GM organism and the environment in which it is to be introduced, and not the method by which it has been produced.

The NRC has further articulated their conclusions into what is now commonly referred to as the 'Concept of Familiarity'. Although familiarity with the characteristics of a particular organism or the environment to which it will be introduced would not necessarily mean it was safe, it can be expected to provide a sufficient amount of information to allow for a judgement to be made of the risks.

For example, familiarity with a new GM plant variety could be established based on comparisons between characteristics of the parent line or other crop species exhibiting similar traits, as well as through the results of actual field tests involving the GM plant. These principles were further elabourated upon by the Organisation for Economic Co-operation and Development (OECD), and as a result, have been referenced in the development of regulatory policies for evaluating the safety of GM crops on a global basis.

More recently, a committee established by the NRC published a report entitled Genetically Modified Pest-Protected Plants: Science and Regulation, based on a review of all scientific and regulatory data collected during the regulatory approval process for GM crops within the United States. The primary objective was to assess independently the effectiveness of existing and proposed regulations for the safety evaluation of GM crops expressing plant pesticides (i.e. plant-incorporated protectants).

No new evidence was identified to suggest plants expressing plant pesticides posed any greater risk to human health or the environment as a result of their genetic modification. In fact, the NRC concluded, 'with careful planning and appropriate regulatory oversight, commercial cultivation of GM plants is not expected to pose higher risks and may pose less risk than other commonly used chemical and biological pest-management techniques'.

However, the NRC report included requests for federal regulatory agencies to further strengthen the current regulatory approval process through better coordination and communication between agencies, on-going investment in the research and monitoring of potential human health (e.g. allergenicity) and environmental impacts (e.g. insect resistance), and by providing greater access to information evaluated in support of regulatory decisions.

## US FEDERAL REGULATIONS FOR AGRICULTURAL BIOTECHNOLOGY

Regulatory systems for the products of agricultural biotechnology have been in existence since the mid- to late 1980s within the United States. The regulatory approach to the safety evaluation of plants developed using rDNA technology has evolved in the best interests of research scientists, industry and the general public. The agricultural products of rDNA technology, such as GM foods and crops, may require approvals from up to three regulatory agencies; the US FDA, the USDA and the US EPA, depending upon the characteristics exhibited by the GM plant, its proposed use and introduced traits. The same standards of safety are applied to all products regardless of the technology used in their development. The US FDA is responsible for ensuring the human safety of all new foods and food components, including products developed using rDNA technology, under the Federal Food, Drug, and Cosmetic Act (I FDCA).

The USDA evaluates the potential of a GM plant to become a plant pest following its environmental introduction under the Federal Plant Pest Act (FPPA). The US EPA evaluates pesticides, including plant systems modified to express pesticides (e.g. insect-protected or virus resistance), under the Federal Insecticide, Fungicide, and Rodenticide Act (FIFRA). As a result, the expression of an insecticidal protein in a food crop would undergo review by the USDA, US EPA and US FDA; a GM food crop exhibiting a modified oil content would be evaluated by the USDA and US FDA; and a non-food horticultural plant developed using rDNA technology for any other purpose (e.g. flower colour) would be subject to review by the USDA alone.

In most instances, obtaining all necessary approvals for the commercialization of an agricultural crop developed using rDNA technology takes a decade or more. However, the exact amount of time required will depend on the need to confirm performance, to evaluate characteristics of the food, environmental effects, and to produce the required amount of seed before the product can be distributed and commercially grown by farmers. Up to five years of field trials (5-10 generations of plants) are required for the developer of a new plant variety to collect sufficient data to meet the reporting requirements of the USDA. An additional five months to two years may be required for the US FDA, USDA and/or US EPA to complete all necessary product consultations, reviews and approvals. Approval for the first commercial planting of a GM food crop was not issued until 1995. Since 1995, more than 40 new agricultural crops developed using rDNA technology have received approval for commercial planting within the United States.

In 1999, approximately 72% of the total 39.9 million hectares (more than 98 million acres) of GM crops grown worldwide were planted in the United States. Herbicide-tolerant soybeans (54%), Bt corn (19%) and herbicide tolerant canola (9%) accounted for approximately 82% of the GM plants cultivated. With an increasing number of agricultural biotechnology products reaching later stages of commercial development, it is anticipated that the overall area planted with GM crops will continue to rise. The regulatory approach to evaluating the human health and environmental safety of GM crops within the United

States is best described as a science-based, case-by-case assessment of hazards and risks. This approach has provided the flexibility required to reduce the regulatory burden placed on products that have been determined to be of low risk or concern. All agencies involved in the regulation of plants developed using rDNA continue to implement and develop policies based on recommendations made within the Coordinated Framework.

In the future, the US FDA, USDA and the US EPA will be dedicating additional resources towards communicating how GM food and food components are regulated within the United States, and how these regulations function to be protective of both human health and the environment.

## FOOD AND DRUG ADMINISTRATION (US FDA)

### Authority

The US FDA is responsible for ensuring the safety and wholesomeness of all food and food components, including the products of rDNA technology, under the FFDCA. The US FDA has the authority for the immediate removal of any product from the market that poses potential risk to public health or that is being sold without all necessary regulatory approvals. As a result, a legal burden is placed on developers and food manufacturers to ensure the commodities utilized and foods available to consumers are safe and in compliance with all legal requirements of the FFDCA.

In order to understand the regulatory approach followed by the US FDA in the safety evaluation of GM crops, it is useful to consider food and food safety from a historical context. People had been consuming foods derived from agricultural crops for many years prior to the existence of any food laws or regulations within the United States. Based on this experience, agricultural crops have been accepted as being safe for consumption as food, without additional testing to demonstrate their safety.

As long as the new crop variety has exhibited similar agronomic properties, and an appropriate taste and appearance, it has been considered safe to consume. As a result, most foods consumed today, in particular whole foods (i.e. fruits, grains and vegetables) and conventional foods, have not been subject to any kind of premarket review or approval by the US FDA. Nevertheless, food scientists have a good understanding that many of the commonly consumed agricultural crops contain natural toxicants (e.g. tomatine in tomatoes, solanine in potatoes, cucurbiticin in cucumber, psoralens in celery, etc.).

As a result, new plant varieties may be subject to routine chemical analyses to ensure that none of these substances is present at potentially harmful levels. This type of general approach has been used in assessing the safety of thousands of new plant varieties that have been developed over a number of decades of crop breeding without compromising the safety of whole foods.

## ROLE

Consistent with recommendations within the Coordinated Framework, the US FDA considered existing provisions of the FFDCA to be sufficient for the regulation of foods and food components developed using rDNA technology. It was concluded that the scientific and regulatory issues posed by the products of rDNA technology were not significantly different from those posed by conventional products.

As a result, GM foods and food components have been subject to the same standards of safety as already exist for the regulation of other foods and food components under the FFDCA. In order to better communicate interpretations of existing provisions of the FFDCA as they relate to the safety evaluation of foods derived from new plant varieties, including the products of rDNA technology, the US FDA released a policy statement in 1992 entitled 'Statement of policy: foods derived from new plant varieties'.

The US FDA has considered the use of genetic modification (i.e. rDNA technology) in the development of new plant varieties to represent a continuum of conventional plant breeding practices (e.g. mutagenesis, hybridization, protoplast fusion, etc.), and as a result, the safety evaluation of all new plant varieties, not just those developed using rDNA technology, have been evaluated based on an objective analysis of the characteristics of a food or its components, and not on its method of production.

## DEFINITION AND SCOPE OF BIOENGINEERED FOODS

The recently proposed rule of the US FDA concerns 'bioengineered foods' which have been defined as 'foods derived from plant varieties that are developed using in vitro manipulations of DNA (generally referred to as rDNA technology)'. As a result, the proposed rule has a much narrower focus than the 1992 US FDA Statement of Policy.

The US FDA has explained the need for a change in emphasis based on their expectations that many of the new plant varieties exhibit a greater potential to 'contain substances that are significantly different from, or that are present in food at a significantly different level than before'. As a result, the substances present in foods and food components derived from new plants developed using rDNA technology are less likely to be considered GRAS, and as a result, will require pre-market approval from the US FDA.

## SAFETY AND NUTRITIONAL EVALUATION

The primary objective of the safety and nutritional evaluation is to demonstrate that the food derived from a new plant variety is as safe or nutritious as foods already consumed as a part of the diet. For new plant varieties, including those developed using rDNA technology, a science-based approach is used to focus the evaluation on the demonstrated characteristics of the food or food component.

The evaluation of a GM food or food component typically involves reviewing information or data on any newly introduced substances, the known levels of toxicants, as well as the nutritional composition of the plant following modification. Substances that raise safety concerns (e.g. toxicants, allergens) would be subject to more extensive evaluation, since both intended and unintended changes may affect the levels of toxicants and nutrients in a food following the modification.

Guidance for performing a safety and nutritional evaluation was provided in the 1992 US FDA Statement of Policy, in a series of flow charts and text that cover:

- The crop that has been modified;
- Source(s) of the introduced genetic material;
- New substances intentionally added to the food as a result of the genetic modification (e.g. proteins, but also fatty acids, and carbohydrates).

Documentation required to support the evaluation typically includes: the purpose or intended technical effect of the modification on the plant, together with a description of the various applications or uses, a molecular characterization of the modification including the identities, sources and functions of introduced genetic material; information on the expressed protein products encoded by introduced genes; information relating to the known or suspected allergenicity and toxicity of any expressed gene products; for foods known to cause allergy, information on whether the endogenous allergens have been altered by the genetic modification; information on the compositional and nutritional characteristics of the foods, including anti-nutrients; and in some instances, comparative results of feeding studies involving the foods derived from plants modified using rDNA technology and the non-modified counterpart. In performing its evaluation, the US FDA is particularly interested in the identification of inherent toxicants, known or potential allergens, assessing the concentration and bioavailability of essential nutrients, the safety and nutritional value of any newly introduced proteins, and the identity, composition, and nutritional value of modified carbohydrates, fats and oils. If additional questions of safety remain following this evaluation, further toxicological studies may need to be performed.

It is recognized that absolute assurance of the safety of any food does not exist. As a result, the goal of the safety evaluation is to establish a reasonable certainty of no harm under anticipated conditions of consumption. With this in mind, experience with the existing food supply has provided the basis for evaluating the safety of new food or food components. Both the Food Advisory Committee and Committee for Veterinary Medicine have been extensively involved in the development of approaches for the safety and nutritional evaluation of foods and food components derived from new plant varieties, including those developed using rDNA technology.

## PRODUCT CHARACTERIZATION

Product characterization takes into consideration information relating to the modified food crop, the introduced genetic material and its expression product, and acceptable levels of inherent plant toxicants and nutrients. All characteristics of the gene insert must be known, including the source(s), size, number of insertion sites, promoter regions, and marker sequences.

It must be established that the transferred genetic material does not come from a pathogenic source, a known source of allergens, or a known toxicant-producing source. The introduced genetic material should be well characterized to ensure that the introduced gene sequences do not encode harmful substances and are stably inserted within the plant genome to minimize any potential opportunity for undesired genetic rearrangement. Analytical data are required to evaluate the nutritional composition, the levels of any known toxicants, anti-nutritional and allergenic substances, and the safety-in-use of antibiotic resistance marker genes.

Any new substances introduced into crops through rDNA technology (e.g. proteins, fatty acids, carbohydrates) will be subject to pre-market review as food additives by the US FDA, unless substantially similar to substances already safely consumed as a part of foods, or that are considered GRAS. To date, substances that have been added to foods through rDNA technology have been previously consumed or have been determined to be substantially similar to substances already consumed as a part of the diet. As such, introduced substances have been considered exempt from the requirement for pre-market approval as food additives with the US FDA. A more rigorous safety evaluation of a GM crop is warranted if the introduced gene sequence(s) has not been fully characterized, the nutritional composition has been significantly altered, antibiotic resistance marker genes have been used during its development, or if an allergenic protein or toxicant has been detected at levels higher than what is typically observed in edible varieties of the same crop species. In any event, determinations as to the safety of substances that have been introduced into new plant varieties through rDNA technology are made on a case-by-case basis.

Although an evaluation of the introduced gene sequences and expression product(s) provides assurance as to their safety, further studies may be required to predict whether unexpected effects may result following their interaction with other genes within the plant. In addition, the product characterization of a GM plant involves assessing sequence homology to known toxicants and allergens, thermal and digestive stability, and if required, the results of both in vitro and in vivo assays to demonstrate lack of toxicity.

## COMPOSITIONAL ANALYSIS

The results of field trials performed over several years serve to characterize the phenotypic and agronomic characteristics exhibited by the plant (e.g. height, colour, leaf orientation, susceptibility to disease, root strength, vigour, fruit or grain size, yield, etc.), as well as to provide the materials required for the compositional analysis. Any anomalies in the phenotypic or agronomic characteristics exhibited by a plant may result in a requirement for additional information. Protein, fat, fibre, starch, amino acid, fatty acid, ash and sugar levels are determined, as well as the levels of anti-nutrients, natural toxicants or known allergens.

Studies of the nutritional composition are performed to determine whether the levels of any key nutrients, vitamins or minerals have been altered as a result of the genetic modification. Based on the results of these studies, a determination is made as to whether the phenotypic and agronomic characteristics of a GM crop or the concentrations of inherent constituents fall within ranges typical of its conventional counterpart.

If inserting a new gene causes no change in any of the assessed parameters, the US FDA can conclude with reasonable assurance that the GM crop is as safe as the conventional crop. If the levels of essential nutrients or inherent toxicants are found to be significantly different in the GM crop, the US FDA may recommend additional action prior to commercialization, such as obtaining food additive status, or the use of specific labels to alert consumers of an altered nutritional content, etc.

## ALLERGENICITY

In consultation with scientific experts in the areas of food safety, food allergy, immunology, biotechnology and diagnostics, the US FDA published guidelines for assessing the allergenicity of GM foods or food components in 1994. The approach to assessment is multi-faceted, incorporating data regarding the origin of the genetic material, and the biochemical, immunological and physicochemical properties of the expressed protein.

The overall assessment is reliant upon the fact that all known food allergens are proteins and, notwithstanding the number of shared properties between allergenic and non-allergenic food proteins, food allergens tend to exhibit a number of similar characteristics. In general, food allergens share a number of common properties: they have a molecular weight of over 10 000 Da; they represent more than 1% of the total protein content of the food; they demonstrate resistance to heat, acid treatment, proteolysis and digestion; and they are recognized by IgE.

For gene sequences derived from known allergenic sources (e.g. peanuts), the developers of GM plants are expected to demonstrate that allergenic proteins have not been introduced into the food. For assessment purposes, it is assumed that any genetic material derived from a known allergenic source will encode for an allergen. To demonstrate otherwise, the amino acid sequence of an expressed protein must be compared with that of known allergens using protein sequence databases.

Furthermore, in vitro and/or in vivo immunologic analyses using the sera of allergic patients sensitive to the source of the genetic material may need to be performed to determine whether or not a potentially allergenic protein is being expressed in the GM food. Some GM foods may be modified to express genes from a source that is not known to be allergenic when consumed.

Under these circumstances, the US FDA follows a similar decision tree-based approach to determining the allergenic potential of the expression product. In assessing these proteins, should any amino acid sequence exhibit homology with a known allergen, the expressed protein would then be evaluated in immunologic tests using the sera of patients known to be allergic to the identified homologous protein.

Regardless of the origin of the genetic material, physicochemical studies are performed in vitro to provide information concerning the expected stability of the expressed protein.

All known food allergens tend to be resistant to digestive degradation, as demonstrated in simulated gastric fluid models, or to decomposition under conditions of food processing. In making a determination regarding a GM food, it is the totality of the biochemical, immunological and physicochemical properties of the introduced protein that provides guidance as to the allergenic potential of such a protein being expressed in food.

The level of protein expressed, produced and consumed as a part of the diet is also a primary indicator of the allergenic potential, since nearly all food allergens are known to be major proteins in their respective foods. If the results of any of these studies suggest an allergenic potential, the US FDA may recommend further scientific evaluation, require special labelling to alert sensitive consumers, or alternatively caution the developer about proceeding with the development of a particular GM food.

Most recently, a joint FAO/WHO Expert Consultation on Allergenicity of Foods Derived from Biotechnology recommended a revised decision tree. This decision-tree strategy was modified from the previous version FAO/WHO to include a revised definition of sequence homology for gene product comparison; a greater emphasis on serum testing, even with gene products without homology to known allergens and not derived from an allergenic source; and animal models to assess potential allergenicity, despite acknowledgement by the Expert Consultation that these models are currently under development and at present, not predictive of food allergies in humans.

More recently, the ad-hoc Open-Ended Working Group on Allergenicity established by the ad-hoc Intergovernmental Codex Task Force on Foods Derived from Biotechnology, considered the FAO/WHO strategy in drafting an approach to assessing the potential allergenicity of foods derived from rDNA plants.

The Codex Working Group recognized the absence of a definitive predicitve test for allergenicity in humans to a newly expressed protein and recommended an integrated, stepwise assessment strategy.The strategy recommended by the Working Group, and accepted by the Codex Task Force for inclusion in the draft forwarded for final adoption by the Codex Alimenarius Commission, is consistent with that of FAO/WHO. However, the Working Group suggested that some of the modifications included in FAO/WHO could contribute to the overall weight of evidence of any conclusion of potential allergenicity (e.g., allergen specific serum depositories, animal models), pending development and validation.

## ANTIBIOTIC RESISTANCE

The use of antibiotic-resistance genes as selectable markers has been common practice in the development of new plant varieties using rDNA technology. Concerns relate to the potential transfer of antibiotic-resistance genes from GM plants to pathogens in the environment or to the gut of humans consuming foods or food components. Issues relating to the use of antibiotic resistance genes were identified in the 1992 US FDA Statement of Policy Statement as well as discussed in additional guidance entitled Guidance for Industry: Use of Antibiotic Resistance Marker Genes in Transgenic Plants.

The guidance provided within these documents was established in consultation with experts in the fields of microbiology, medicine, food safety, bacterial and mycotic diseases, and includes suggestions with respect to the continued safe use of antibiotic-resistance marker genes by the developers of new plant varieties. The use of marker genes that encode resistance to clinically important antibiotics has raised questions as to whether their presence in food could reduce the effectiveness of oral doses of the antibiotic or whether the gene present in the DNA could be transferred to pathogenic microbes, rendering them resistant to treatment with the antibiotic.

The risk of transfer of antibiotic-resistance genes from plants to microorganisms considered to be pathogenic to humans, however, is considered to be minimal if not insignificant. Furthermore, the potential risks are becoming less of a concern as more developers are beginning to research the use of alternative technologies (e.g. non-resistance-based markers) in plant breeding. The conclusions with respect to the safe use of antibiotic-resistance marker genes are consistent with the findings of other national and international food safety organizations.

### Consultation and Filing Process

The submission of a Pre-market Biotechnology Notification (PBN) has recently been proposed as a mandatory requirement for the commercialization of bioengineered foods and food ingredients within the United States. Minimal differences exist between the information to be submitted as a part of a PBN and that previously presented in voluntary consultations with the US FDA.

As proposed, developers are still being encouraged to consult with the US FDA as early and as often as necessary in the development of a bioengineered food, such that any potential scientific or regulatory concerns can be identified and addressed prior to the submission of the PBN. Guidelines for performing consultations with the US FDA were released in a publication entitled Guidance On Consultation Procedures: Foods Derived from New Plant Varieties in 1997.

The publication recommended an approach for developers to proceed by submitting a request for consultation, outlined the internal process by which all requests would be handled, and included additional guidance as to the type of safety and nutritional information to be presented during consultation with the US FDA. Once sufficient safety and nutritional information has accumulated to demonstrate that a product is safe and in compliance with the FFDCA, developers typically schedule a consultation to present their scientific findings and conclusions to the US FDA.

Consultations prior to notification not only serve to keep the US FDA informed of advances made in the application of rDNA technology in food production, but also keep the developers of bioengineered foods aware of emerging safety, nutritional or regulatory concerns of the US FDA. Consultations are considered complete when all safety and regulatory concerns between the US FDA and the developer have been resolved.

The US FDA proposes to perform an initial evaluation of a PBN within 15 days of receipt to determine completeness, at which point, if considered complete, the PBN will be filed, and a response can be expected within 120 days.

*The US FDA does not issue a product approval per se, but informs the developer by letter that*:

- The evaluation period has been extended;
- The notice is not complete and why;
- It has no further questions 'at this time' based on the information that has been presented.

## Labelling

The FFDCA defines what information must be disclosed to consumers on a food label, such as the common or usual name, and other limitations concerning the representations or claims that can be made or suggested about a food product. All foods must be labelled truthfully and not be misleading to consumers.

Taking this into consideration, the FFDCA does not stipulate the disclosure of information on the basis of consumer desire to know. Labelling may be considered misleading if it fails to reveal material facts in light of representations that are made with respect to a product. The labelling of foods derived from new plant varieties, including plants developed using rDNA technology, was originally addressed in the 1992 US FDA Statement of Policy, and most recently discussed in Draft Guidance for Industry for the voluntary labelling of bioengineered foods.

To date, the US FDA is not aware of any information that would distinguish foods developed using rDNA technology (e.g. bioengineered foods) as a class from foods developed through other methods of conventional plant breeding, and as such, have not considered the method of development a material fact requiring disclosure on product labels.

Nevertheless, after extensive consultation, including thousands of written comments and a series of public meetings, the US FDA has observed 'a general agreement that providing more information to consumers about bioengineered foods would be useful'.

## Requirements for Labelling

Special labelling is required if the composition of the bioengineered food differs significantly from its conventional counterpart. For example, for a food that has been genetically modified to contain a new major sweetener, a new common or usual name or other labelling may be required.

Similarly, if a GM food contains an allergen that consumers would not expect to be present in that food, special labelling may be necessary to alert sensitive consumers. If a protein commonly associated with an allergic reaction (e.g. peanut protein) is transferred to another food through genetic modification, the US FDA would evaluate whether labelling would provide sufficient consumer protection. If labelling would not be considered to provide a sufficient level of protection, the US FDA would take appropriate steps to ensure the GM food would not be marketed.

Therefore, current policy requires a GM food to be labelled when the resulting product poses a safety issue or is substantially different from its conventional counterpart, and as a result, could be considered to pose a misrepresentation to consumers. The 1992 US FDA Statement of Policy and the recent Draft Guidelines do not consider the use of rDNA technology in the development of food products to be a material fact requiring specific disclosure on the label.

Rather it is a method of development, similar to other methods of plant breeding, which have not required disclosure on the label. Bioengineered foods cannot be distinguished compositionally from foods modified through more conventional methods and thus do not require specific disclosure through labelling. The Draft Guidelines reaffirm the US FDA position that bioengineered foods do not require special labelling.

**Voluntary Labelling**

To provide guiding principles for voluntary labelling, in recognition of the desire of certain manufacturers to label foods as produced either with or without bioengineering, the US FDA published a 'Draft Guidance for Industry: voluntary labelling indicating whether the foods have or have not been developed using bioengineering'.

Emphasizing that the use of rDNA technology 'is not a material fact', the US FDA recognizes that some consumers want disclosure of bioengineered content and that some manufacturers wish to provide it. In response, Draft Guidance was issued with suggestions concerning the use of labelling statements that are not considered misleading.

## PRESENCE OR USE OF rDNA

The US FDA provided several examples of how disclosure of bioengineering can be accomplished, be informative and not be misleading.

- Example 1 'Genetically engineered' or 'This product contains corn meal that was produced using biotechnology'. These disclosures reveal the minimum amount of optional information about bioengineering.
- Example 2 'This product contains high oleic acid soybean oil from soybeans developed using biotechnology to decrease the amount of saturated fat.' This statement explains the proper and required name of this type of soybean oil, since it differs from what would be considered as soybean oil. The optional comments about biotechnology and decreasing saturated fat provide information that could be seen as benefits of the product.
- Example 3 'These tomatoes were genetically engineered to improve texture.' This example was one included to illustrate how it could be misleading to consumers if they cannot discern a difference in texture, but would not be misleading if they can tell a difference. If the former, and the new texture is to facilitate processing, then this intention should be made clear, i.e. 'to improve texture for processing'.
- Example 4 'Some of our growers plant tomato seeds that were developed through biotechnology to increase crop yield.' This is another example of optional information that would explain an indirect, agricultural benefit.

## ABSENCE OR NOT BIOENGINEERED

The US FDA provides an important commentary in the Draft Guidance about 'genetically modified organisms (GMO)' versus 'bioengineered'. It would be technically inaccurate to use the phrase 'not genetically modified' or 'GMO-free' to mean that bioengineering was not used. This is because most conventional foods have been genetically modified over the years by traditional crop breeding practices.

*Examples of acceptable voluntary statements would include*:

- We do not use ingredients that were produced by biotechnology.'
- This oil is made from soybeans that were not genetically engineered.'
- Our tomato growers do not plant seeds developed using biotechnology.'

*Another important point is made about a term such as 'GMO-free' that is misleading for two reasons*:

- Many foods do not contain organisms anyway and therefore should not be labelled as 'organism-free';
- Free' implies complete absence or 'zero' amount of bioengineered material. In a practical sense, it is impossible to demonstrate analytically the complete absence of anything. Therefore, in reality, a threshold for possible adventitious presence of a low level of bioengineered material may be necessary and has been the subject of much debate.

*The Agency also provides additional guidance about misleading statements that*:

- Could be interpreted to suggest that the absence of bioengineering would make the food superior to the bioengineered alternative;
- Claim the absence of one bioengineered ingredient when the food contains another ingredient that is bioengineered;
- Claim that a food is not bioengineered when in fact this type of food (e.g. green beans) has never been modified through rDNA technology.

## DEPARTMENT OF AGRICULTURE

The Animal and Plant Health Inspection Service (APHIS) of the USDA is responsible for the protection of domestic agricultural resources and the environment from the threats posed by pests and diseases. The Federal Plant Pest Act (FPPA) has been the primary statute adapted by APHIS in regulating the products of agricultural biotechnology. Under the FPPA, APHIS has the authority to restrict the import of plant pests, to introduce measures (e.g. quarantines) to prevent their movement, or to destroy plant pests cultivated in violation of the FPPA. The requirements under the FPPA are equally applicable to scientists within company, academic, research, and private organizations.

### Definition of Plant Pests

A plant pest is any agent, substance or organism, or parts thereof, that may cause injury, disease or damage, either to plants or to the environment in which it is introduced. APHIS maintains a list of organisms considered to be plant pests, and therefore, subject to regulation under the FPPA. If an organism is not on the list, it may still be subject to regulation as a plant pest if there is reason to believe it is or will act as a plant pest.

Organisms considered plant pests, or that exhibit the potential to be plant pests, are treated as 'regulated articles' under the FPPA. The regulations extend to the introduction of GM organisms meeting the definition of a plant pest or for which there is reason to believe they are plant pests. APHIS considers the environmental release of a GM crop to be equivalent to the introduction of new organisms. Therefore, until proven otherwise, a GM crop is considered a 'regulated article' under the FPPA.

The use of the term 'plant pest' in reference to GM plants means the 'non-pest' nature of the crop has yet to be determined. To date, all field tests conducted have demonstrated that GM crops exhibit no more 'pest-like' traits than their conventional counterparts. All GM plants developed to date have involved the use of at least one designated 'plant pest' as either promoters or vectors, and as a result, have been subject to regulatory review under the FPPA.

APHIS takes the position that an entire plant can be designated as a regulated article, even if it was not developed using plant pests, if the plant is the product of genetic engineering, and the agency determines or has reason to believe it is a plant pest. Given this position, APHIS would likely challenge any attempt to introduce a GM plant into commerce unless the developer has first gone through the APHIS regulatory review process.

### Requirements for Commercialization

A number of years of field testing are required to evaluate the agronomic and product quality characteristics exhibited by new plant varieties developed within a labouratory or greenhouse. APHIS regulations outline procedures for obtaining a permit or providing notification prior to the importation, interstate movement or release of a regulated article, which of course, includes field testing. Prior to an introduction, a proponent must provide notification to APHIS of its intentions or submit an application to obtain a permit. The notification and permit requirements have evolved as an extension of the long-standing programme for the regulation of plant pests within the United States. Since APHIS has issued permits or acknowledged the receipt of notifications for the field release of more than 6500 new crop varieties.

## Permits

For APHIS approval, developers must demonstrate that a new plant variety poses no significant risk to other plants in the environment or is at least as safe as similar crop varieties. APHIS review and approval are required for the shipment of seed and for the conduct of field trials involving new crop varieties. APHIS issues site-specific permits for field tests or releases into the environment.

Prior to field testing, APHIS may seek further clarification of study objectives, specify how, when or where research may be conducted, establish the data to be generated, collected and reported, stipulate additional requirements for the monitoring or securing of test sites, and specify methods for the disposal of crop residues that remain at the end of a study.

The results of field trials are often used in making a determination as to whether a GM plant exhibits any effects on non-target species, or poses any unique plant pest problems. For field trials conducted under either notification or permit, developers of GM plant varieties are required to submit reports to APHIS that include: details regarding the methods of observation; the data collected; and interpretations concerning the effects on plants, non-target organisms, or the environment.

In the event that a GM crop is likely to be the subject of a petition for a determination of non-regulated status, developers must provide a description of known and potential differences, and substantiate that the regulated article is not likely to pose a greater plant risk than the organism from which it was derived. An environmental assessment is a requirement of the NEPA, Council on Environmental Quality regulations, and USDA procedures for issuing permits under the FPPA. Under the permit process, developers of GM plant varieties must disclose information concerning the plant, test facilities, and control measures in place for its transport and field testing. Based on this information, the USDA performs an assessment of the potential environmental impact following a release. If the agency reaches a 'Finding of No Significant Impact' (FONSI), a permit may then be issued.

### Notifications

Depending on the plant species and intended introduction, applying for a permit from APHIS can be a complicated and time-consuming process. In the spring of 1993, APHIS introduced a simplified notification process as an alternative to applying for a permit, which in most instances applies to the introduction of GM plants. To qualify for notification, a GM plant must be introduced in accordance with the eligibility criteria.

Originally, six GM crops (corn, cotton, potatoes, soybean, tobacco and tomatoes) qualified for notification; however, the process was more recently expanded to include most other crop species not capable of becoming a noxious weed. The developers of new GM plant varieties are encouraged to contact APHIS when making a determination as to whether a GM plant would qualify for notification. If a particular GM plant does not qualify for notification, a more involved regular permit process must be followed.

For a GM plant to qualify for notification, it and its predecessor must not be a noxious weed, the introduced gene sequence(s) must be stably integrated within the host, not originate from a plant or animal pathogen, exhibit a known function that will contribute to plant disease, or encode for substances that could be potentially toxic or infective to non-target organisms, such as pharmaceuticals, or viral components other than those coat proteins already established as safe.

Performance standards also exist for the conduct of field trials to ensure that an adequate level of containment is provided for all introductions. It is the responsibility of the developer to ensure that an adequate level of containment is provided for field trials involving GM plants under notification.

The performance standards utilized will vary according to the biology of the plant and the nature of the introduction. In general, this includes specifying methods for the proper handling, shipment, field monitoring and disposal of plant material at the end of the study.

As a part of the notification process, an applicant must provide information about the plant, identify the source of any genes used; the method of genetic modification; and the size, date and location of any proposed field test or introduction. Notification is required at least 10 days prior to the interstate movement or 30 days in advance of the field testing or importation of a regulated article that qualifies for notification with APHIS.

**Determination of Non-regulated Status**

Once sufficient data have accumulated through labouratory and field trials performed under permit or notification, a developer may petition APHIS for a determination of non-regulated status. A DONRS allows a GM plant to be grown, tested or used for crop breeding without further regulatory oversight by APHIS. APHIS will grant a DONRS if sufficient evidence can be provided to demonstrate that no plant pest potential exists (i.e. that the GM plant is not expected to become a pest, poses no significant risk to the environment, or is as safe as conventional plant varieties).

The developer is required to demonstrate a lack of change in disease or pest resistance status, an absence of any potential for contributing to development of a new plant pathogen or pest, as well as to address any questions relating to potential environmental consequences of the release. APHIS examines several parameters when making a determination as to whether a GM crop should be considered a plant pest. Reporting requirements include submitting a detailed rationale for development; overview of the biology of the crop (competitiveness, survivability, dormancy); molecular characterization; protein expression; morphologi-cal/phenotypic characteristics; outcrossing/geneflow; weediness; insect and disease susceptibility; and, recombination potential (for viral genes).

Other considerations include assessing whether a GM crop could itself become a weed. In making this determination, factors that are considered include: how the seeds are dispersed in the environment; whether the seeds are capable of surviving over the winter; the potential for the development of volunteer plants; and whether these plants could reproduce into viable offspring.

Another important consideration is whether the seeds or plants are capable of surviving outside of a managed agricultural environment (e.g. watering, fertilizer, etc.), and how the introduced traits could influence the viability of the crop under marginal conditions. APHIS also takes into consideration potential adverse effects on wildlife, including birds, beneficial insects and mammals, following the introduction of a GM crop. Some of this information comes from field trials that are conducted in multiple locations for several years.

By observing crops growing under actual conditions of use in the field, scientists can compare insect populations existing within a field planted with GM crops to those coexisting in fields planted with a non-modified variety of the same plant. Field trials are also useful in identifying any changes in the plant physiology, such as height, colour, leaf placement, time of flowering, etc., that result from modifications in the genome of the plant.

Monitoring of these changes during field trials allows APHIS to evaluate how these changes might benefit or adversely impact the behaviour of wildlife in contact with these crops. Knowing that wildlife, such as deer, often feed on agricultural crops, APHIS also takes into consideration the nutritional content of the GM crop. Comparisons of the essential nutrient content are often made between the GM plant and its conventional counterpart.

Other interests may include determining the impact of the modification on the levels of adverse factors (e.g. natural toxicants or anti-nutrients) present in the plant. APHIS maintains a list of all crops that have received a DONRS. At any time, APHIS retains the authority to prohibit the commercial planting of a GM crop if it is determined that it is becoming a plant pest.

## PETITION PROCESS

Before a GM plant can be freely transported and commercialized, the developer must first petition the APHIS for a DONRS. A petition for a DONRS requires the submission of extensive data on the introduced gene construct, effects on plant biology and effects on the ecosystem, including the spread of the gene to other crops or wild relatives. Developing all data necessary to support a petition for a DONRS may take months or years depending upon the scientific issues that need to be addressed.

Based on the information provided within the petition, APHIS evaluates the potential impact (e.g. toxicological) that the introduced modifications will have on non-target organisms in the environment, including threatened or endangered species. A petition for a DONRS must provide a rationale for the development and introduction of the GM crop, starting with genotypic and phenotypic information on both the host and donor organism(s).

This includes a detailed description of the methods of transformation, identifying all gene sequences (e.g. promoters, leader sequences, introns, selectable markers, etc.), their function, and potential for plant risk following introduction. A combination of different analytical methods may be used to demonstrate the stable integration of an introduced gene sequence within the genome, its inheritance and expression in host plants. To the extent possible, the gene(s) of interest (e.g. resistance, marker genes), levels of expression, and im pacts on levels of inherent plant toxicants in plant tissues, under experimental and actual growth conditions in the field, must be characterized. Field performance studies (e.g. leaf morphology, pollen viability, seed germination, viability, insect susceptibilities, disease resistance, yield, etc.) are used to identify any differences or similarities in phenotypic characteristics exhibited by the GM and non-modified crop. The extent of field performance data required in support of a petition for non-regulated status depends not only on the nature of introduced gene sequences, but also on the physiology of the host plant. Less extensive field performance studies are required for plants that are: highly domesticated (e.g. corn); self-pollinating (e.g. soybean); male sterile; those with high seed germination rates (> 90%); or unlikely to influence their potential weediness or fitness (e.g. delayed ripening, oil seed content).

Whereas, GM crops exhibiting a tolerance or resistance to cold, salt, biotic (e.g. insects, pathogenic agents), or other abiotic (e.g. herbicide) stresses, are subject to more extensive field performance testing. APHIS considers several parameters in reaching a DONRS, including whether the GM plant can cross-pollinate with other plants in the wild, and if it can, the ecological consequences (e.g. insect resistance, herbicide tolerance) that might result.

Extensive databases are maintained on species capable of cross-pollinating with GM crops (e.g. corn, soybean, cotton, canola, etc.), based on the results of years of breeding experiments, biological surveys and research conducted by agricultural and weed specialists within APHIS. In instances where out-crossing might be expected to occur, APHIS may stipulate the planting of refuge areas to limit the potential for adverse ecological effects following commercial planting.

## AGENCY RESPONSE

APHIS will grant a determination of non-regulated status if the developer can sufficiently demonstrate a lack of plant risk. In making a determination that a GM plant will no longer be regulated, APHIS prepares two documents, an environmental assessment and a DONRS to satisfy regulatory requirements under NEPA and FPPA, respectively.

Both documents are developed based on the review of data submitted as a part of the petition for a DONRS. Once the DONRS has been granted by APHIS, a GM crop no longer requires notification or permit prior to its movement or release within the United States. A review for DONRS is generally completed within 10 months of submitting a formal application to the APHIS, allowing sufficient time

products of rDNA technology that act as pesticides. Leading up to this publication, the US EPA had acquired considerable expertise in the registration of biological pesticides (e.g. microbial).

The US EPA was first involved in discussions regarding the potential risks of pesticides developed using rDNA technology in the early 1980s, during a series of public meetings involving the FIFRA Scientific Advisory Panel (SAP) and the Biotechnology Scientific Advisory Committee (BSAC). It was recognized that the potential risks posed by plants modified to express pesticidal components through rDNA technology were unique and warranted additional consideration relative to conventional pesticides.

For example, the potential for introduced genetic material to out-cross from a modified plant to a sexually compatible wild or weedy relative through pollen spread was one of the unique, theoretical risks posed by plant-pesticides. As a result, the EPA proposed a plant-pesticide rule in 1994, outlining their interests in the regulation of products.

A package of three proposed final rules was released early in 2001, redefining regulations for plant-incorporated protectants (PIPS), previously referred to as plant-pesticides. Essential elements of the proposed rules for the regulation of GM plants expressing pesticidal traits have been followed since 1994. The final rules provide further clarification of how PIPS will be regulated under FIFRA and FFDCA, including specific exemptions and became effective in September

## Definition of Plant-incorporated Protectants (PIPS) or plant-pesticides

US EPA's final rule clarifies the regulation of whole plants that contain PIPS. When plants are used intentionally for controlling pests in some way, such uses meet the definition of a pesticide under FIFRA. Nevertheless, the agency exempts the plants themselves from regulation, focussing instead on the substances they contain, previously referred to as plant-pesticides and now known as PIPS.

A plant-incorporated protectant (PIP) includes the substance produced by the living plant, and all genetic material necessary for its production (including promoters, enhancers, etc.). Thus, genes and substances (e.g. enzymes) responsible for the conversion of natural plant constituents into pesticidal substances would be considered PIPS.

To date, PIPS registered by the US EPA have been proteins and the genes required to make these proteins within the plant. Crops involved include potatoes, cotton, field corn, sweet corn and popcorn. Clearly, Bt toxins represent the most predominant class of PIPS that have been registered by the US EPA. Major reasons for this include the success of microbial products containing Bt and their long history of use for nearly 40 years. It has been estimated that combined, Bt corn, potato and cotton were cultivated on approximately 10 million acres in 1997, 20 million in 1998, and 29 million in 1999. As a result, less chemical insecticide has been applied, and there have been significantly higher crop yields. In addition, due to reduced opportunity for opportunistic fungal infections of insect-damaged corn, lower levels of mycotoxins have been produced, reducing the toxicological risk to both humans.

## Exemptions

US EPA has concluded that living plants with pesticidal activity do not require a high level of regulatory scrutiny by the agency and thus, are exempt from regulation under FIFRA. The agency commented that focussing on the substance produced within the plant, i.e. the PIP eliminates any need for the US EPA to register plants, thus conserving agency resources and reducing potential overlap with other agencies, notably the USDA. For example, although whole plants that are used as biological control agents themselves, such as chrysanthemums, are exempt, substances that are extracted from plants, such as the insecticidal material pyrethrum, are not excluded from regulation under FIFRA.

Further exemptions have been proposed, based on familiarity and presence of the pesticidal substances in the food supply. Only one of the additional proposed exemptions has been included in the final rule, and all others have been proposed for further public comment. Within the final rule, EPA has exempted PIPS and encoding genetic material originating from plants sexually compatible through conventional breeding but not through rDNA technology. The rationale for this exemption is that such substances could be bred, either naturally or through conventional plant breeding, from close plant relatives that share genetic material from a common gene pool. On the other hand, genetic modifications that were never before possible can be made using rDNA technology, involving the genes from sexually incompatible plant species.

Therefore, the exemption of plant-pesticides developed using rDNA technology, involving genes from sexually compatible plants has not been finalized as a part of the supplementary proposal to the final rules for PIPS. Other exemptions for further comment have also been identified within the supplemental proposal. The first group are PIPS that 'act primarily by affecting the plant'. The group have in common some sort of defence mechanism such that a pest would be less able to attach to, penetrate, or successfully invade a plant.

*Examples include*:

- Some sort of structural barrier such as wax or plant hairs;
- Inactivation of or resistance to a pest or substance (e.g. toxin) produced by a pest;
- Creation of a deficiency in a nutrient or growth factor required by a pest;
- Hypersensitivity response that would contain or limit spread of the pest (e.g. necrosis of surrounding plant tissue, creating a functional barrier);
- Plant hormones that are naturally occurring but with change can affect a plant in a variety of ways with potentially dramatic results.

The other classes of PIPS proposed for exemption include viral proteins and their encoding genes. The expression of genes for viral proteins within a host plant confers resistance to infection by that virus as well as certain related viruses. These PIPS are plant virus specific, and as such, would not be expected to pose any risk to the public.

### Herbicide-tolerant Crops

In the case of herbicide-tolerant crops, EPA determines whether the property of increased resistance may encourage higher rates or more extensive application of an agricultural chemical to food crops. The higher exposures to herbicides would be subject to evaluation through risk assessment, requiring the submission and review of detailed data. If the risk is determined to be acceptable, a herbicide product would require label extensions for the addition of new tolerant crop varieties.

### Experimental Use Permit (EUP)

Small-scale research involving PIPS is covered under the permit and notification process that exists for new plant varieties administered by APHIS. The US EPA requires that an Experimental Use Permit (EUP) be obtained for all research and field testing involving greater than 10 acres of a plant modified to express a pesticidal substance. The EUP process allows a developer to gather the product use, performance, residue and other types of data, required in support of the registration of PIPS. All US EPA decisions concerning applications for EUPs are published within the Federal Register for comment.

### Registration Molecular Makers

For the registration of a plant-incorporated protectant, the EPA evaluates the potential impact, on human health and the environment, of the pesticide substance and of the genetic material necessary for its expression within the modified plant. With this in mind, the EPA developed their guidance for evaluating the safety of PIPS based on data requirements that were established for microbial pesticides.

A number of modifications to the reporting requirements were required to focus the assessment on non-target toxicity rather than pathogenicity, as well as to evaluate the expression of the pesticide substance within the plant itself, rather than by spraying, etc. Although specific reporting requirements for the registration of PIPS are established on a case-by-case basis by the EPA, the general categories of information and data required in support of the registration of a plant-pesticide include:

- Product characterization
- Toxicology
- Environmental fate and exposure
- Non-target organism effects
- If necessary, plans for pest resistance management.

The EPA continues to hold meetings of their SAP to re-evaluate the reporting requirements that have been established for PIPS and to ensure that all potential human health and environmental concerns are being addressed as new scientific information becomes available. In addition, the developers of PIPS are encouraged to consult with US EPA during all phases of the research and development process, particularly when it concerns an environmental release.

As a part of their review, scientific evaluators within the EPA weigh the potential hazards inherent to the generic consideration of PIPS and the potential for exposure to the protectant substance. The characterization of possible hazards, in the context of exposure, allows for estimates of potential environmental risk to be established.

The commercial approval of a PIP may take between 15 and 18 months following receipt of a full data package by the US EPA, and it involves a public notification and consultation process. However, this time frame would be considered an ideal, since in reality, deciding what constitutes a 'full' data package may take several rounds of discussions with the US EPA.

## PRODUCT CHARACTERIZATION

Product characterization allows for consideration of any potential hazard(s) that might be associated with anticipated exposure to a plant-incorporated protectant. This process includes a review of the source of introduced genetic material, including all regulatory sequences, the modifications that have been made, methods of transformation, inheritance, stability and expression in the host plant. For the introduced pesticidal substance, the anticipated modes of action, specificity and toxicity to both target and non-target organisms are evaluated. Target organisms may include weeds, insects and diseases (viral, bacterial, and fungal) affecting a crop. The crop itself, as a volunteer weed, may also be of concern following crop rotations. Once a PIP and any potential for exposure have been characterized, the data required in support of an environmental and human health risk assessment can be defined.

## ECOLOGICAL EFFECTS

Environmental fate and exposure are important considerations when evaluating the potential impact of PIPS on the environment or on non-target organisms (e.g. wildlife, beneficial insects). Depending on the crop and introduced pesticide substance, data on the level of expression and rate of degradation may be required, first for the tissues of the modified plant, and secondly, the fate in soil, water, or any other environmental compartment.

Such data allow the potential for exposure to the pesticide substance to be evaluated for non-target organisms via feeding on, or through decomposition of, the modified plants. Studies are arranged in tiers, starting with Tier I, consisting of acute toxicology and basic characterization studies that help predict environmental fate. Depending on the results obtained in Tier I, and the level of concern, additional studies may be requested from higher tiers. It is expected that most PIPS will only undergo Tier I testing, since negative results are expected at relatively high exposure levels.

The 'maximum hazard approach' when used in Tier I provides a high level of assurance that adverse effects would not be expected to occur under actual conditions of use within the field. If the results of Tier I reveal a potential hazard at doses within the range of expected concentrations in the environment, additional studies would be requested in higher tiers to better characterize the potential for any kind of effect. Data requirements may be satisfied by generating test data or by requests for data waivers with credible justification.

The type and extent of dietary studies required will depend on properties of the crop and on the introduced pesticide substance. Dietary studies typically involve feeding plant tissues (e.g. grain, pollen) that express the PIP or that have been spiked with the pesticidal substance. However, when determined necessary, the pure pesticidal substance may be administered to non-target species at doses of up to 10 to 100 times the expected level of exposure in the field. Dependent upon the potential for exposure, additional studies may also be requested involving any number of other non-target organisms (e.g. honeybees, green lacewing, ladybird beetles, parasitic wasps, earthworms, etc.). When toxicity is observed, the potential for exposure becomes an important consideration when ascertaining whether an adverse effect might be expected under actual conditions of use in the field. The requirement for provision of a resistance management plan is not routine practise for PIPS, but depends on an evaluation of the potential for resistance to develop and the threat to existing or functionally equivalent pest management practises.

To date, resistance management plans have only been placed on Bt PIPS. If considered appropriate, the EPA may request that a resistance management plan be submitted as a part of a condition for registration. A resistance management plan may require a commitment towards the generation of additional research data, to implement structured refuges, to perform annual resistance monitoring, to initiate remedial action plans, and to educate growers. The EPA has worked closely with academia, other federal agencies, public interest groups, industry, and growers, to establish resistance management plans based on the most current science.

## HUMAN HEALTH EFFECTS

For agricultural food crops, special consideration is given to evaluating the safety of the pesticidal substance introduced as a component of consumed foods. Rodents (often mice, to conserve test material) are orally administered the pesticide substance at doses of up to 100 000 times the levels of expected human consumption as a part of the diet.

Although the modified plant tissues would be considered the most appropriate vehicle for determining the acute toxicity of a PIP, it is not typically feasible, since only small amounts of the pesticidal substance are present in plant materials.

Therefore, an alternative source (e.g. microbial) is often used to generate a sufficient amount of identical pesticide substance for testing purposes. In this case, the equivalency of the test substance obtained from an alternative source to the PIP (e.g. composition, identity, etc.) being expressed within the plant is an important consideration that must be well documented.

For pesticide substances that are proteins, the principal concern relates to the potential for the introduction of a new allergen or toxin into the food supply. Since most allergenic food proteins are stable to digestion, in vitro digestibility studies are used to predict how long a

scientific data. For example, the SAP for Bt Plant Pesticides met three times in the year 2000 to consider issues related to the safety of plants that had been genetically modified to express insect-specific Bt toxins. The US EPA released a decision to temporarily extend registrations for all currently registered pest protected plants throughout the 2001 growing season.

Additional provisions, such as the recently strengthened resistance management requirements for plant-pesticides, along with the original registration conditions, were considered to provide adequate protection during the extended time period. In addition, the reassessment process for these products has been designed to assure maximum transparency and opportunity for public comment in the regulatory decision-making process.

## CONCLUDING COMMENTS

The US FDA has been praised for its science-based regulatory approach to evaluating the safety of products developed using rDNA technology based on the characteristics of the food product, and not the means by which it has been created. On the other hand, the regulatory approach of both the USDA and US EPA has been criticized for discriminating against the products of rDNA technology.

A consortium of 11 major international scientific societies and the Council on Agricultural Science and Technology (CAST), an organization representing 36 scientific and professional groups, have strongly criticized the US EPA for their proposed approach to regulating plant-pesticides on the basis of the process used in its development rather than the characteristics exhibited by the plant.

More recently, the US Congressional Committee on Biotechnology released a report entitled 'Seeds of Opportunity' recommending revisions to the USDA and US EPA approaches to the regulation of GM plants. It was felt that the proposals by both agencies failed to take into consideration the current scientific consensus on the potential risks associated with agricultural biotechnology.

Overall, the report was supportive of the science-based policy of regulation, based on the characteristics of a food product and not on the means by which it was created. As a part of their role, each federal regulatory agency involved in assessing the human health and environmental safety of GM crops attempts to ensure that only the most current scientific data, information, and methods are referenced.

The US FDA consults a Food Advisory Committee, the US EPA refers questions to Scientific Advisory Panels, and the USDA has recently established its own advisory panel. Aside from periodic reviews by credible independent research organizations such as NAS, the individual policies of each agency have been subject to regular review for harmonization with approaches recommended by international standard-setting bodies, such as OECD, WHO and FAO. It is expected that many of the future food products of rDNA technology in agriculture may not be equivalent in composition to their conventional counterparts.

However, the existing decision-tree approach to regulation anticipates these types of improved food products, and provides for the case-by-case assessment of their safety. Field testing and pre-market review for food safety provide the required level of assurance that the foods derived from the application of rDNA technology in agriculture are at least as safe as existing foods, and are consistent with all existing standards of food safety.

All new varieties of vegetables, fruit, corn and soybeans developed using rDNA technology have been subject to considerable scientific and regulatory review by the US FDA, the USDA and/or the US EPA. As such, the developers of new crop varieties have had to take into consideration not only the interests of farmers, food processors and consumers, but also the expectations of regulatory authorities.

Both regulators and consumers have had a number of opportunities to influence the acceptance of a new crop variety, and continue to have significant influence beyond final regulatory approval and commercial introduction. It is expected that the issues of human health and environmental safety will continue to be of major concern in the future as a greater number of GM food products reach later stages of commercial development.

A recent trend has been observed towards the commercialization of products that offer more transparent benefits and value to consumers. The new products of rDNA technology will be expected to further challenge the flexibility that has been exhibited by the US regulatory process. As a result, both industry and regulatory authorities will need to be responsive to uphold the human health and safety record achieved for GM product introductions. The vast majority of the new products of rDNA technology are not expected to be commercially available until after 2005.

The crops and foods improved through rDNA technology have been developed with much more precision, and their human health and environmental safety has been assessed in more depth and detail than any other crops and foods developed using more conventional means. They are routinely evaluated for any potential impact on both the environment and food safety. Despite years of experience and research, no verifiable data implicating rDNA technology as a food safety or environmental hazard have presented themselves. The research and development of GM crops have been performed under extremely controlled conditions and tight regulatory oversight from research and development within the labouratory through field trials and commercial planting for food-use.

Based on evaluations conducted to date, no evidence exists to suggest that GM foods currently sold on the market pose any significant human health concerns or that they are in any way less safe than foods derived from crops developed through more conventional means. Although the foods developed using rDNA technology have been subject to rigorous scientific review for safety, the US regulatory system has allowed for considerable advancement in the application of the technology, and significant product innovation.

To continue to be successful, it is recognized that the US FDA, USDA and US EPA will need to cooperate: the approval process must be clear, timely and transparent; emphasis must be on science-based/risk-based regulatory assessments and decisions; the regulatory process must be flexible to adapt to new information; and, the agencies must be staffed by full-time, highly qualified scientists in many diverse fields.

Over the past 25 years, the approach to regulating the products of agricultural biotechnology has evolved in response to the experience gained during the research and development of applications for rDNA technology. In the future, as the number of products developed using rDNA technology continues to increase, additional research will be needed to better evaluate any potential effects on human health and the environment, and to further refine the scientific basis of making regulatory decisions.

Within the United States, the regulatory approach to evaluating the safety of products developed using rDNA technology continues to evolve as new applications continue to emerge and our understanding of the potential risks improves. To date, existing regulations have proven adequate to assure the safety of all new products developed using rDNA technology, and additional innovation has not been discouraged. Accordingly, the National Research Council has recommended that any new rules for the regulation of GM plants be sufficiently flexible enough to reflect improvements in scientific understanding.

In the future, as rDNA technology finds broader application, and the new prospects for this technology become commercial realities, the regulatory approaches to evaluating the safety of GM food crops are expected to evolve accordingly.

## Chapter 3

# Transgenic Food Crops

The application of recombinant DNA technology has revolutionized plant breeding in recent years. Genetic transformation in plant breeding greatly increases the gene pool and broadens the scope of genetic changes that modern breeders may draw upon. The first generation of genetically modified (GM) crop plants hold great promise in carrying new or improved agronomical traits that require less intensive farming methods or which have improved quality traits.

These GM crop plants have been produced to improve farming production systems, aiming, for instance, to reduce negative impacts on the environment, such as fertilizers, herbicides and pesticides. In general they are of little direct interest to the consumer. However, the genetic manipulation of the primary and secondary metabolism of plants offers distinct possibilities in the development of nutritionally improved products, raw materials with added pharmaceutical value, and crops with real health benefits for the consumer.

These are second-generation GM plants. Recently, several prototypes in molecular plant breeding with novel traits for (non-) food and industrial applications have been introduced into the market, and many more GM crops have the prospect of commercial exploitation in the near future. The safety and wholesomeness of these GM crop plants raises questions beyond those posed by conventional foods. In this context, traditional plant breeding of most food crops has conformed to the standard 'generally recognized as safe' (GRAS). In some instances safety is assumed to be assured during performance and quality testing by breeders (e.g. glucosinolates and erucic acid in rape seed).

However, the process is largely focused on the comparison of agronomic and phenotypic characteristics, such as colour, maturity, yield or disease resistance. Products with an unusual taste or that have harmful effects following ingestion or skin contact have normally been rejected from the breeding programme. However, the safety of conventional plant varieties has usually been presumed from the assumption that prudent consumers will avoid those species once it is known that they cannot consume them without adverse health effects. When the commercial production of GM plants started, it sparked a debate centred on issues associated with concerns on environmental, human and animal safety matters. Many people questioned whether the 'simple' GRAS approach, based on historical and empirical experiences, is sufficient to guarantee food safety of the GM crop plants. It is also asked whether the risks of chronic exposure of humans and animals to GM crops are adequately covered, and whether large-scale breeding of GM plants will have detrimental effects on ecosystems.

## FOOD SAFETY ISSUES

Although the variety registration of existing crop plants has not resulted in adverse effects in humans, the conventional assessments carried out by plant breeders was not considered to be sufficient to ensure food safety of GM crop plants.

As a result, Europe issued a regulation on novel foods and novel food ingredients that sets the legal framework for the market introduction of genetically modified organisms (GMOs). This came into force. The Novel Foods Regulation establishes a system of mandatory pre-market approvals of novel foodstuffs, including GM plants.

The accompanying EU guideline gave a clear indication of the types of data that are needed to form the basis of an assessment. In particular, four central safety issues have been set forth related to food and feed safety:

- The nutritional and toxicological consequences of inserted gene products;
- The potential of pleiotropic (unintended) effects in the host organism due to the insertion event;
- The allergenicity of expressed proteins and novel foodstuffs;
- The potential of gene transfer to human and animal gut flora.

This means that in transgenic insect-resistant Bt tomatoes encoding the C-terminal truncated Bt2 gene derived from a Bacillus thuringiensis (Bt) strain, IAb5, the novel protein has to be evaluated on its own merit, while the safety and wholesomeness of the 'remaining' transgenic tomato is assessed separately. Whereas, for example, in antisense RNA exogalactanase tomatoes, which have improved rheologic characteristics due to downregulation of the endogenous exogalactanase activity, the safety assessment is primarily focused on the 'remaining' novel fruit. In general, there is basic agreement on the safety issues to be addressed. One particular area of concern is the possibility of unexpected or unintended metabolic perturbations due to genetic modification that may alter, for instance, levels of nutrients and health-influencing components. The use of rDNA techniques does not necessarily result in fundamental changes in crop plants compared to the food produced by conventional breeding methods. However, it should be emphasized that a uniform international agreement for the evaluation of GMOs is urgently needed.

For example, an agreement on how to establish in practical terms the similarities or differences between the 'remaining' GM crop plant and the traditionally bred plant seems to be much more difficult to achieve. This chapter is devoted to the description of advanced strategies for the evaluation of potential unintended effects in GM crop plants. Pleiotropic (unintended) alterations in agronomic traits or composition may arise from insertion mutagenesis or as a result of metabolic effects of the novel gene product(s).

The challenge is to gain the ability to discriminate between metabolic alterations due to somaclonal variations, unintended effects or natural diversity. Thereto, emphasis is given to a platform of technologies designed to identify changes in the molecular machinery of crop plants. As is the case for detection methods for GM crops and derived products, there are needs and criteria for sampling strategies, statistical models and manufacturing standard reference matehers recently.

## UNINTENDED EFFECTS

Strategies for assessing the food safety and wholesomeness of novel foodstuffs are currently in the exploration phase. Above all there is a need for uniform and harmonized quantitative methods for the testing of potential unintended effects in whole foods, such as GM crop plants are. The safety of, for example, synthetic chemicals and food additives is established by assessing separate elements of the compound in question.

Although very successful in case of single chemicals, the evaluation of a whole food product may not be a simple addition of individual assessments of the constituents. Traditional testing protocols involving labouratory animal 90-day feeding trials with (whole) food products are far from ideal. This type of animal feeding study with complex food matrices is complicated by the likelihood of nutritional imbalances leading to dietary problems, confounding factors, an insensitivity for specific endpoints, and the impossibility of using large safety margins, if any. Therefore, in a global context several regulatory, research and industry-based bodies have proposed alternative concepts for the safety evaluation of genetically engineered foods and food ingredients. In particular the Organisation for Economic Co-operation and

Development (OECD), the Food and Agriculture Organization (FAO) and the World Health Organization (WHO) have developed concepts and principles for the safety evaluation of GMOs and derived products.

Their reports concluded that risk assessments should be directed at demonstrating that a GM crop plant or derived food product is as safe as a traditional or previously authorized product. Accordingly the EU Competent Authority incorporated these general principles into their Novel Foods Regulation and accompanying guidelines. In this context, novel foods and food ingredients that are considered to be 'no longer equivalent' to traditionally bred plants are subject to labelling as defined in the Council Regulation.

## Substantial Equivalence

The term 'substantial equivalence' appears in the EU Novel Foods Regulation as a major principle in the risk assessment. It emphasizes that, within the context of the Regulation the evaluation of the safety and nutritional value of a GM plant should be focused on a comparative analysis with conventionally bred products. OECD's Group of National Experts on Safety in Biotechnology enunciated this comparative approach towards the safety assessment of transgenic food plants as the concept of substantial equivalence.

In their report, it is assumed that existing food organisms possess a long history of safe use (i.e. GRAS). Consequently, the most practical approach to the determination of safety is to consider whether the GM plants are substantially equivalent to analogous food product(s), if such exist. Once substantial equivalence of a novel foodstuff has been established, it provides assurance of safety that is equal to or better than that of its comparators.

Further clarification of the concept of substantial equivalence originates from a WHO sponsored workshop. It has been concluded that the establishment of substantial equivalence is not a safety assessment per se, but a dynamic, analytical exercise. Thus, it may mean that analysis of, for instance, gene expression patterns, global changes in protein expression and/or differences in metabolic capabilities (the metabolome) of the novel product in comparison with conventionally bred products should be performed in order to examine equivalence.

This strategy has worked satisfactorily for the assessment of the first-generation crop plants for which sufficient background knowledge is available. However, relatively less experience has been gained with the safety and wholesomeness evaluation of novel food plants for which no substantial equivalence can be established. Even in the case of single gene modification, potential alterations in metabolic pathways re difficult to identify and, consequently, the implications of the enetic modification process on the metabolism of plants are poorly nderstood. Moreover, data on the mechanisms by which plants egulate, for instance, their response to environmental stress factors nd other conditions are remarkably scanty.

## :ritical Nutrients and key Toxicants

Application of the concept of substantial equivalence appears in ractice to lead to various interpretations. Different requirements for sk assessment with regards to the potential of unintended effects exist s a result of genetic modification. Actual approaches entail a onsideration of the characteristics of host and donor organism, henotypic properties and toxicological evaluation and include omparative determinations of any changes in critical (anti-)nutrients nd key toxicants for the food source in question.

In general, the analysis of an expanded spectrum of components ; unnecessary, but should be considered if there is an indication from ther traits that there may be an unintended secondary effect of genetic odification. However, compositional analyses based on single arameters using a list of crop-specific critical (anti-)nutrients and key oxicants as a minimum requirement of screening for possible ariations in the content has its limitations.

Limited data are typically available for the less important food omponents, as in less well-known crops most critical (anti-)nutrients nd natural toxicants will be unknown. Moreover, the toxicants to be ssessed may be partly influenced by the cellular function of the roduct encoded by the inserted gene. Furthermore, the natural ariation of critical comparators may mask the evidence of secondary ffects due to genetic modification.

Thus, there is no criterion that determines whether a difference etween a GM plant and its natural counterpart is outside the literature alues for the ranges of parameters that express natural diversity. elevant information on indicators of unexpected effects may not have een observed during the development of an edible transgenic crop pecies. In the future other compounds could become of importance, uch as anti-oxidative, oestrogenic and anti-carcinogenic compounds.

In determining critical nutrients and/or key toxicants, differences consumption patterns and practices of processing and consumption various geographical settings and cultures must also be recognized. : is our opinion that conclusions about relative safety and holesomeness based on a list of selected single components may not e equally valid in a global context and in all times. Therefore, there is an urgent need to set up generic approaches using platform technologies to evaluate undesirable metabolic perturbations in GM food crops.

## A Post-genome Challenge

In our view the identification of unintended effects encompasses the ability to interpret and use innovations in molecular genetics research such as genomics, proteomics and metabolomics on a crop-by-crop basis. This will establish a whole data package directed towards a holistic view on possible unintended side effects due to genetic modification. In essence, the focus is shifting towards molecular characterization to understand functional activity.

Vital to this approach will be informative profiles of molecules at different integration levels, e.g. mRNAs, proteins including post-translational modifications and molecules at the level of primary and secondary plant products and metabolites that might be useful benchmarks for the detection of unintended effects. The relative importance of these various 'profiling' techniques in establishing data sets for assessing unintended effects is not indicated by their order and will vary from species to species.

Alterations in, for example, expression levels, post-transitional modification, interactions or chemical composition do not per se imply that the product is less safe or even unsafe. Such a platform of technologies may determine whether unintended effects exist between the GM plant and its control(s) or whether there is substantial equivalence apart from certain well-defined expected differences.

Measures of the relative activities of various molecular constituents associated with metabolic capabilities in non- and modified plant cells under different conditions will provide new insights into how metabolism and genetic modification are orchestrated. But it should be emphasized that the fields of plant genomics and proteomics are still in their infancy.

## mRNA FINGERPRINTING

A very promising strategy to evaluate the equivalence of genetically modified crops is gene expression profiling. Conventional methods for the analysis of differential gene expression include Northern blotting, S1 nuclease protection, comparative expressed sequence tag (EST) sequencing, differential display and serial analysis of gene expression. More recently, Kok *et al.* have designed a method of detecting altered gene expression by means of mRNA fingerprinting or reverse transcription polymerase chain reaction (RT-PCR). Based on the original concept of Liang and Pardee specific subsets of the mRNA population of a tomato plant were amplified. This enables the expression of different genes in the GM crop and genes in the parental line to be compared and any empirical information about significant alterations to be established.

The method is still under development and requires further validation, but has been shown to be reproducible and able to detect important differences in gene expression. Some of the most significant remaining problems are that the method largely depends on individual skills based on experience and is time-consuming, even in its most simplified form.

## cDNA MICROARRAYS

Over the past few years, the DNA microarray technology has emerged as a powerful high-throughput method for the analysis of gene expression. At present, microarrays for gene expression studies generally consist of cDNAs, which are physically deposited onto small glass surfaces. The major advantage of DNA microarray technology over conventional gene-profiling techniques is that it allows small-scale analysis of expression of a large number of genes in a sensitive and quantitative manner.

Furthermore, it allows comparison of gene expression profiles under a large number of different conditions. The cDNA microarray technology has already been successfully introduced in disciplines ranging from cell and developmental biology to drug development and pharmacogenomics. In the following a number of technical aspects of the technology as well as its potential value for the safety assessment of genetic modification of food plants will be discussed.

The cDNA microarrays are produced by deposition onto a solid surface of purified PCR products corresponding to specific genes. Typically a few nanolitres of DNA solution (100-500 μg/ml) are spotted using a micro-dispensing robot. Essential information on robot construction and protocols can be found on the Internet. In general, with respect to the spotting devices, a higher density of spots is accomplished when using a passive dispenser (> 2500 DNA spots/cm$^2$).

Thereto, glass slides covered with a positively charged layer (e.g. poly-l-lysine) or carrying reactive groups are most commonly used. To analyse the expression of genes in a plant cell or tissue sample, polyA+ RNA (typically 0.5-2.5 μg is required for each reaction) is purified and labelled by reverse transcription using an oligo-dT primer incorporating a fluorescently labelled nucleotide. The labelled cDNA pool is then hybridized to the microarray. The required hybridization conditions (i.e. sample concentration, stringency of hybridization) an parameters, such as detection level and the correlation betwee transcript concentration and hybridization have been determined fo each experimental setting.

The intensity of a hybridization signal, which is a measure of th relative expression level of each gene, can be read by charge-couple device (CCD) cameras or confocal/non-confocal laser scanners. RNA extracted from two different samples can also be simultaneousl analysed with a microarray by labelling each with a differer fluorescent label using either Cy-3 or Cy-5. At RIKILT-DLO we ar currently testing whether the analysis of differential gene expressio using DNA microarrays is an informative strategy for the safet evaluation of GM plants.

To address this question, DNA libraries are constructed an enriched by a subtractive cloning approach for cDNAs preferentiall expressed in either green or red tomato fruit. These subtracted cDNA as well as control cDNAs representing known tomato genes (i. sequence information derived from public sequence databases), hav been spotted on arrays and are used for comparison of gene expressio profiling of control and genetically modified tomatoes.

In the near future we will construct more defined arrays containin genes of toxicologically relevant pathways, such as those involved i the synthesis of natural plant toxins (e.g. the glycoalkaloids α-tomatin and α-solanine).

## PROTEOME PROFILING

The introduction of Bt genes of bacterial origin may theoreticall influence the nature of post-translation modification (e.g. glycosylatio in the transgenic tomato fruit. This is because as well as the impact the protein moiety, for example, the type of glycosylation pattern ma alter due to inactivation of glycosidases and glycosyltransferases i the host cell. In addition, glycosylation of the inserted Bt gene produ in the plant might result in altered protein activity and stability.

The importance of structurally elucidating these oligosaccharide especially the N-glycans, can be attributed to their widesprea functions in the cell. These functions include correct folding, biologic activity and stability of proteins as well as involvement in pla development. Moreover, N-glycans represent antigenic epitopes b themselves. Immunological as well as biochemical studies revealed th allergenic potency for the α1,3-fucosylation of the proximal N acetylglucosamine residue. Additionally, also the α1,2-xylosylation the α core mannose is suspected to be an allergenic epitope. Recentl there is increasing evidence that these N-linked oligosaccharides may have a substantial impact on the immunogenicity of glycoproteins, as N-glycans are suspected to play a certain role in the adverse food reactions of hypersensitized patients.

At the Institute for Agrobiotechnology, Austria, this issue was tackled by developing a generic technology for isolating, purifying and characterizing N-linked oligosaccharides in plant tissues. Our low-temperature acetone powder method appeared to be most suitable for the isolation of (glyco-)proteins from plant tissue (e.g. tomato fruit). The resulting (glyco-)protein fraction was proteolytically digested and corresponding glycopeptides purified by cation-exchange chromatography.

The oligosaccharides (N-glycans including α1,3-fucosylated oligosaccharides) were released from the peptide backbones by N-glycosidase A (PNGase A) and fluorescently labelled with 2-aminopyridine. Subsequently, pyridylaminated glycans were separated and analysed structurally by repeated two-dimensional high-performance liquid chromatography (2D HPLC) using a size-fractionation and a reverse-phase column in combination with an exoglycosidase treatment.

MALDI-TOF (matrix-assisted laser desorption ionization time-of-light) mass spectrometry was applied for further confirmation of the observed structures. Sixteen different N-glycosidic structures could be detected in tomato fruit. The two most abundant glycans in all the tomato samples showed identical properties to those of the major N-linked oligosaccharides of horseradish peroxidase (MMXF: Man(3)Xyl-FucGlcNAc(2)) and pineapple stem bromelain (MOXF: Man(2)Xyl-FucGlcNAc(2)), respectively, and accounted for about 65-78% of the total glycan content.

Oligomannosidic glycans occurred in only small quantities (3-9%). The majority of the N-glycans found was α 1,2-xylosylated and carried an α1,3-fucose residue linked to terminal N-acetylglucosamine. A structural element was identified and shown to be an IgE-reactive determinant, which contributes to cross-reactivities among non-related glycoproteins, and is present in a wide variety of plants extracts.

It is proposed that this structural element might also be of importance in some cases of food and pollen allergy. The carbohydrate profiling of, for example, populations of green and red-ripe fruit of non-modified and modified Bt tomato plants (expressing the cryIA5 gene) revealed no differences in the nature of N-glycans, nor were there significant differences in the relative amounts of glycan. Interestingly, the relative amounts found were quite similar across seasons and comparable to the commercial variety Notoro used as an extended control. As far as the N-linked oligosaccharides were involved it was not possible to detect significant alterations in the á1,3-fucose content of Bt tomatoes due to genetic modification.

## PROFILING OF THE METABOLOME

Among other factors, identity of a complex GM crop plant with its traditionally bred parent entails metabolism and the level of (un-)desirable substances contained therein. Therefore, we would like to build up an overall picture of components associated with different metabolic capabilities in the (non-)modified plant tissue. A multi-compositional analysis of the so-called metabolome may overcome the test limitations of, for example, single critical nutrient or key toxicant analyses and problems related to their limits of natural diversity.

An attractive alternative is the utility of high-resolution proton (1H) nuclear magnetic resonance (NMR) spectroscopy (1H-NMR) in combination with separation techniques (liquid chromatography (LC)). Recently, we have shown that the approach of chemical fingerprinting using off-line LC-NMR can obtain information on possible changes in complex plant matrices due to environmental effects as long as GLP-like conditions for sample handling, data acquisition and automation of data handling are maintained. In principle, it will be possible to screen all the different low molecular weight components (MW <10 kDa) present in the metabolome of the plant tissue.

## CHEMICAL FINGERPRINTING

The principles for establishing a chemical fingerprint were directed towards the detection of alterations in compositions in 1H-NMR spectra arriving from two populations, for example, non-modified counterpart(s) and an engineered one. In this way the unprocessed parts of the plant used for human consumption were fractionated and analysed, i.e. in tomato comparison on the fruit level.

Genetically modified tomato varieties, such as the antisense RNA exogalactanase fruit, have been studied using this innovative technology. The spread of measurements of the individual profiles was comparable for the GMO and its controls. By subtraction of the 1H-NMR spectra the differences of the means were calculated at 99% confidence intervals between the GM crop average (n = 8/batch per line) and its isogenic control average (n = 8/batch per line). The difference of the means was defined here as all differences of the means of the amplitudes of individual NMR signals obtained from spectra of various populations. The differences of the means were used to quantify how the transgenic crop plant differed from its parental line. The normalized 99% confidence intervals of the normalized means of fractions A-E after eightfold independent replicates demonstrated that differences in amplitudes exceeding at least 20% could normally be detected. This is an adequate sensitivity to allow statistical evaluation, and is in line with recommendations of the Nordic Council. They proposed that if the average value of a parameter differs by more than 20% an explanation should be sought.

With the purpose of establishing those differences that may arise from metabolic effects linked to genetic modification the default statistical analysis used a mixed effects randomized block model. Random block effects such as variations in the stage of cultivation, location, logistics or climate had a significant impact on the overall chemical fingerprints of cultivars. For example, chemical fingerprints varied much more between field sites and seasons than between replicates in one plot.

A tomato background mean (i.e. crop mean value) was established by constructing an extended range of non-modified tomato references for comparison with natural variability, and to assess the impact of external factors (i.e. variations in season, climate, etc.). The crop mean value was used to quantify how the differences of the means between the transgenic crop plant and its control differed from the natural diversity of commercial counterparts.

To differentiate between compositional changes either due to genetic modification, genetic variability or environmental variations the chemical fingerprints were studied in a hierarchical approach.

*Hence, the fingerprints of the GMO are compared to those of:*

- The isogenic control line grown side-by-side under identical conditions in one plot;
- The isogenic control line grown side-by-side under identical conditions at multiple sites;
- An extended range of commercial varieties of that crop;
- The influence of downstream processing.

It was recognized that 45% of all compounds of the control line harvested in different years varied considerably in concentration. Most differences were found in fractions A, C and D. A similar degree of impact on composition was observed in the case of processing temperatures (e.g. 100°C versus 70°C) on tomato juice. These findings were compatible with the protease activity surviving the 70°C treatment. But in fruit harvested from various individual plants, divided into two populations, approximately 95% of all constituents were practically identical in concentration. Also the determination of the differences of the means between transformant tEG to its isogenic control bred side-by-side in the same plot showed that 249 out of 3000 amplitudes (8%) varied in concentration.

Assignments by NMR indicated that, for example, citric acid was present in fraction A of the control at 1.4 times the levels found in the transformant. However, in fraction A of the transformant glutamic acid and/or glutamine showed an increase of 1.3 times the levels found in the isogenic control.

On the other hand, one single aromatic compound, á-lycopene, appeared to be present in fraction E of transformant tEG at 2.5 times the concentrations found in the parental line. No alterations were observed in ß-carotene levels. These effects may be the result of the antisense downregulation of the tEG1A gene in tomato. In our view the natural diversity must be taken into consideration, however, when interpreting the biological relevance of any significant difference found in the differences of the means between the GMO and its parental line. The parental line was extended by entering non-isogenic commercial references (i.e. crop mean value).

Although limited in the range of lines, the comparison of the crop mean value to the differences of the means showed that the contribution and magnitude of differences (e.g. citric acid, glutamic acid/glutamine, á-lycopene) diluted out. It was recognized that in the agricultural practice the conditions of culturing, processing and storage of appropriate references will be mostly unknown and they may have a non-isogenic genotype. Notwithstanding these complicating random block effects we recommend that, for example, the chemical fingerprint should not be limited to the transgene and its isogenic parent bred under identical circumstances only, but should also include a sampling through multiple sites and a comparison with an extended range of controls of that crop.

Otherwise it would be a too limited approach towards the establishment of the claim of substantial equivalence and it would neglect the ranges of natural variations in that crop. In future, if many more lines have to be compared with each other this approach may benefit from multivariate statistical methods.

## TOXICOLOGICAL PROFILING

The significance of detected compositional alterations may need to be further explored by in vitro toxicity assays in order to screen for potential adverse effects and for mechanistic studies. It may even be that the latter approach leads to a redesign, refinement or even replacement of (sub-)chronic rodent feeding trials. In vitro systems derived from organs and tissues from animals and humans, and various types of cultured recombinant cell lines, are successfully used for screening for the toxic potential of single compounds.

However, their ability to screen for the potential toxicity of whole foods or extracts thereof has been insufficiently explored up till now, and has been examined by us within the framework of a tiered safety evaluation of GM plants. The validity of using in vitro test assays for comparative testing of whole food products, food ingredients and extracts thereof depends on the sensitivity and selectivity of the toxic response induced upon exposure. However, since these models are not yet validated, they should be considered as early warning or alert-systems to identify potential changes in toxicity of the engineered product only.

## GENOTOXICITY ENDPOINTS

The mucosa of the gastrointestinal tract is clearly a primary, potential site of action for the effects of novel foods. Therefore effort has been concentrated on the development of a battery of intestinal cell systems (e.g. IEC-6, IEC-18, Caco-2 and INT407 cell lines) with different endpoints for initial cyto- and genotoxicity testing. The choice of a suitable method of sample preparation was considered of utmost importance in assaying vegetables.

Freeze-dried tomato fruit was extracted (10% suspensions, w/v) in water, 0.9% saline or chloroform:methanol (2:1, v/v) immediately before incorporation into the culture medium. It was encouraging that aqueous and chloroform/methanol extracts of, for example, red-ripe tomatoes up to a maximum of 10% suspensions (w/v), had little or no general toxicity once the pH (> 6.0) and osmotic pressure had been taken into account.

No toxicity to the intestinal cells Caco-2 and IEC-18 under short-term conditions was detected in red-ripe, non- and antisense RNA exogalactanase tomato using alpha-glutathione S-transferase (α-GST) lactate dehydroyrenase (LDH) leakage, reduction of 3-(4,5-dimethylthiazol-2-yl)-2,5 diphenyltetrazolium bromide (MTT conversion), neutral red (NR) uptake and total cellular protein as in vitro endpoints.

On the other hand, green tomato fruit displayed a pronounced toxic and cell-damaging effect. No statistical significant differences were found between transformants and isogenic control lines in terms of their activity in the assays. Treatment of tomato tissue with pepsin (ratio pepsin to tomato protein 1:250, w/w) did not result in an increase of or loss of inherent cytotoxicity. Supplementary investigations with naturally occurring plant components, for example, quercetin, α-carotene, α-lycopene, genistein showed that only steroid glycoalkaloids, such as α-tomatine, exhibited a rapid toxic effect, whereas α-tomatidine was ineffective. It was noted that the in vitro models did show general matrix-related toxic effects.

For example, α-tomatine diluted in a matrix of extracted red-ripe tomato tissue (final concentration of 5-10% w/v) showed an increase of dose-response-dependent cytotoxicity in human Caco-2 cells (e.g. IC50 7.9 ± 2.1 mg/L versus IC50 33.9 ± 5.3 mg/L of α-tomatine only). Using the COMET assay, suspensions of freeze-dried tomato fruit up to a maximum of 10% (w/v) did not exhibit DNA-damaging effects in a range of gastrointestinal cells derived from rats (IEC-6 and IEC-18) and humans (Caco-2, INT407).

Genetic modification did not result in increased genotoxicity, but fruit extracts were able to suppress the DNA-damaging effects of the known genotoxins $H_2O_2$ and N-methyl-$N_2$-nitro-N-nitrosoguanidine (MNNG). Although these endpoints are important toxicological measures they do not indicate the molecular mechanisms of toxic damage. So far, pronounced adverse effects could only be observed with glycoalkaloids. The screening for subtle differences in toxic responses of other plant constituents appeared to be limited. However, therefore, the use of critical endpoints for such studies other than cytotoxicity should be clearly defined.

## ASSESSING TRANSCRIPTION FACTORS

There is a demand for in vitro models that not only determine the (geno-) toxic potency of plant components or complex mixtures, but also provide information on the mechanism by which that compound exerts its toxic effect. As an in vitro profiling system the eukaryotic stress gene assay (i.e. CAT-Tox(L)) was assessed. This utilizes human HepG2 cells stably transfected with chloramphenicol acetyltransferase (CAT) reporter constructs.

The aim of the CAT-Tox(L) assay is to collect and quantify molecular responses to cellular stress and toxicity that have a transcriptional component in their regulation. The assay employs an ELISA method to measure the transcriptional activity of the promoter or response element constructs. The advantage is that many of the stress responses occur before any measurable cytotoxicity, thus allowing monitoring of stress pathways at sublethal levels.

Since 14 unique transcription responses can be measured simultaneously, the CAT-Tox(L) provides comprehensive cellular stress profiles. The assay generates specific stress gene induction profiles for a variety of stressful and toxic plant components such as genistein, diadzein and sinigrin. Aqueous extracts of red-ripe, non- and antisense RNA exogalactanase tomato fruit dosed up to 0.05 g per assay did not cause any cytotoxicity to the HepG2 transfectants nor were there molecular responses to cellular stress or toxicity. Extracts of non-modified and modified green tomato fruit showed cytotoxicity and induced the construct containing the xenobiotic response element (XRE).

This gene-induction profile was not comparable to that of, for example, the reference molecules dioxin, 3-methylcholanthrene, benzo[a]pyrene or α-napthoflavone. These polycyclic aromatic hydrocarbons were characterized with a co-induction of cytochrome P450 1A1 and glutathione-S-transferase Ya subunit- CAT fusion constructs. There was substantial evidence that the systemic fungicides fenarimol and bupirimate did not induce any of the genes in this assay. There was no promotor in the assay that could be linked to universal cellular stress.

A-tomatine caused severe cytotoxicity at high concentrations and yet did not induce any of the genes (i.e. causing membrane damage). However, it was observed that α-tomatine diluted in a matrix of red-ripe tomato tissue (final extract concentration 5% w/v) elicited an increased cytotoxicity concomitant with a dose-dependent response profile involving XHF (collagenase involved in inflammatory reactions sensitive to mitogens and cytokine IL-1). So far, the approach of the CAT-Tox(L) model shows potential but there is a need for validation based on molecular mechanisms of plant constituents.

Although the induction of these stress genes represents cellular perturbations, the mechanistic validations need to be extended before any conclusions can be drawn on the value of these findings. Certainly, the development of a database to explore relationships between a wide variety of compounds stressful to plants and meaningful stress gene profiles may serve as molecular fingerprints unique for a particular component.

## GENE EXPRESSION PROFILING

A second potential application of DNA microarray technology in safety evaluations concerns screening for biological effects of (fractions of) plant compounds that have been found by other profiling techniques to be present or modified in the genetically modified crop plant when compared to the parental line and other references. At RIKILT-DLO we are testing the feasibility of this strategy by analysing defined food compounds for possible effects on human intestinal gene expression. To this end, human intestinal cell lines are exposed for different time periods to various amounts of crop plant components and the mRNA isolated. The labelled mRNA is hybridized to DNA microarrays containing unknown intestine-specific genes, as well as already identified control genes known to be intestine specific or to be functional indicators (for toxicological processes, apoptosis, etc.).

As a source for unknown, intestine-specific genes we use cDNA libraries from human intestine biopsies and human intestinal cell lines which have been enriched by a subtractive cloning approach. Having established the usefulness of this approach for the identification of biological functions of defined plant compounds, these cDNA microarrays will be exploited to screen, for example, GM crop plants for possible effects on human health. The concept of substantial equivalence is broadly accepted as a basis for the risk assessment of GM crops and derived novel foods.

Data requirements for establishing substantial equivalence should be based on identifying unintended metabolic perturbations in the engineered crops that have been caused by genetic modification. A generic approach cannot be envisaged, however. In all cases the collation of data should be directed by the type of genetic modification and related consequences. It is recommended that unintended effects should be identified on a case-by-case basis, since such effects are dependent on the genotypic and phenotypic parameters of the new gene products involved.

The application of innovations in molecular genetics research will help to define the conditions under which the new food products can be marketed. The present approach has generated a sound scientific basis for the evaluation of the potential of unintended effects in GM crops and derived food products. The results may assist regulators and legislators to test and, if necessary, to improve the applicable regulations.

Moreover, the technologies described and the results obtained may serve as a general framework for the risk assessment of GM crop plants, and may contribute to a better understanding by the public of recombinant DNA techniques in plant breeding and their implications. It is concluded that a parallel approach focusing on a comparative analysis using informative profiles of molecules at various integration levels (e.g. mRNA, protein, metabolites) including a toxicological profiling of relevant plant matrices, offers good prospects for the identification of hazards related to unintended effects. The choice of comparators and of external parameters to support a claim that a GM crop plant presents no unintended effects compared with its traditional counterpart should be based on sound scientific judgement. It is recommended that data are generated on the unprocessed part of the plant, such as the tomato fruit or potato tuber, used for human and/or animal consumption.

Furthermore, data banks should be set up containing information on natural variations of essential plant or other product constituents (i.e. crop mean values). The chemical fingerprinting technique, using a combination of off-line liquid chromatography and proton-NMR imaging, appears to be a powerful screening method for the detection of secondary effects in the metabolome that may be applied on a routine basis.

The carbohydrate profiling technique is useful for the detection of unexpected changes at N-glycan levels; in particular it is applicable to the detection of post-translational modifications in newly expressed proteins as well as in the whole GM crop plant. DNA microarray technology is rapidly evolving and much effort is being put into improving spot density, reduced production time, and increased reproducibility and sensitivity.

Improving the latter is critical because it makes it possible to use smaller amounts of starting material. It will depend on technical aspects such as the quality of the scanners and spotting machines, the development of fluorescent dyes with improved characteristics (e.g. narrow excitation and emission peaks; high level of photon emission; resistance to photo-bleaching), supports with reduced background and more target sequence binding capacity.

DNA microarrays generate a huge amount of complex hybridization data and major challenges are to develop more advanced computer software to help to find statistically significant correlations within, and between different experiments, and to link the data to sequence information and submitted expression profiles available in public databases. Application of these technologies in combination with other measurements of, for instance, the performance and quality may limit or even replace animal feeding studies aimed at the detection of unintended effects.

Together, these techniques, known as functional genomics, will enable us to answer questions about what happens to expression and protein composition when a plant cell changes metabolic state, for example, due to genetic modification or environmental factors. It is foreseen that this holistic view on the cellular machinery will be achieved soon, as the coincidence of genome sequencing with improvements in the analysis of expressed proteins is of growing importance and continues to be successful.

Eventually, microarrays for genomic studies will probably also be used to evaluate the numerous constituents of GM crop plants. Some constituents might not cause obvious changes in cellular behaviour or morphologic appearance but could cause subtle metabolic alterations that would show up when the mRNA content was interrogated by an array. It is likely that microarrays might be used to evaluate the life cycle of a plant much more precisely and to understand the complex metabolic control systems of plants. The ultimate test for GM crop plants will be acceptance by the public in large.

It is therefore extremely important that industry, the scientific community and the regulatory authorities are able to respond to queries from the public by providing transparent information on the criteria for safety assessment. Improved safety assessment at the molecular level should refine and complement strategies by anticipating the potential risks and sources of unintended effects in GM plants.

## Chapter 4

# Virul Disease Resistant Crops

## AGRICULTURAL IMPORTANCE

Squash belongs to the Cucurbitaceae, a family of crops that is an important source of food throughout the world. While not as important as the cereals or oilseed crops, such as soybean, nor other vegetable crops such as potatoes and tomatoes, cucurbits are nevertheless grown on significant acreages worldwide.

In 1998, cantaloupes and other melons were grown on 1 044 672 hectares, producing 17 764 188 metric tonnes of food; cucumbers and gherkins were grown on 1 567 389 hectares, producing 26 673 943 metric tonnes; while pumpkins, squash, and gourds were grown on 1 136 083 hectares, producing 14 169 983 metric tonnes. Cucurbit crops, including squash, are used for food in all regions of the world, but are also used for oil and other non-food uses, such as musical instruments or household implements.

### IMPACT OF VIRUS DISEASE

Squash, like the other major cucurbit crops, is susceptible to infection by several viruses. The most damaging are cucumber mosaic virus (CMV), papaya ringspot virus watermelon strain (PRSV-W), watermelon mosaic virus 2 (WMV2) and zucchini yellow mosaic virus (ZYMV). While the extent to which viruses cause economic damage in cucurbits, and squash in particular, is not well documented, the problem is extensive enough that diligent effort to breed squash resistant to CMV, PRSV-W, WMV2 and ZYMV has been a goal of breeders for many decades.

#### Disease breeding in Squash

Sources of genes for resistance or tolerance to viruses have been reported in various relatives of squash and in Cucurbita pepo itself. However, introducing useful levels of resistance to more than one virus into commercially acceptable cultivars has been difficult to achieve.

While releases of zucchini-type and English marrow-type varieties carrying resistance to a single virus (CMV) has been achieved, breeding for multiple virus resistance has only recently been successful. A summer squash cultivar, Whitaker, has recently been developed by researchers at Cornell University.

This cultivar is resistant to multiple pathogens, including viruses and fungi. The difficulty of this accomplishment is emphasized by the fact that Whitaker was the result of work conducted by a team of researchers for more than a decade. Because the sources of resistance were related species not normally sexually compatible with C. pepo, these researchers had to employ a commonly used breeding tool, embryo rescue, to overcome the barriers to crossing between species.

## SQUASH AS A SPECIES

Cucurbita pepo is a crop native to North America. It consists of several subspecies and varieties that are completely interfertile, and are therefore presently considered part of a single species. C. pepo ssp. pepo includes cultivated pumpkins, marrows and a few ornamental gourds; C. pepo ssp. ovifera var. ovifera includes cultivated crookneck, scallop and acorn squashes, and most ornamental gourd cultivars; C. pepo ssp. ovifera var. texana includes free-living populations found in Texas; C. pepo ssp. ovifera var. ozarkana includes free-living populations found in the Mississippi Valley and the Ozark Plateau; C. pepo ssp. fraterna includes non-cultivated populations found in northeastern Mexico.

Populations of C. pepo exist today under non-cultivated conditions in the United States. These free-living populations range from northeastern Mexico and Texas, to Alabama, and through the Mississippi Valley to Illinois. Some free-living populations, especially in Arkansas, Louisiana and Mississippi, occur as weeds in soybean and cotton fields, and are difficult to control.

Gene exchange between free-living and cultivated populations has been documented. Furthermore, some free-living populations may have originated as escapes from cultivation, which might have undergone subsequent introgression with other nearby cultivated, or other free-living populations.

### Genetic Engineering of Squash

Because of the need to obtain useful virus resistance in a commercially acceptable cultivar, an alternative to breeding was attempted: the direct introduction of genes conferring resistance into already known commercial genotypes. The genetic engineering approach was to introduce the coat protein genes of the viruses in question, a strategy first successful in tobacco, and subsequently shown to be effective in cucurbit species. Constructs containing viral coat protein genes were introduced into proprietary inbred lines of C. pepo ssp. ovifera var. ovifera via Agrobacterium-mediated transformation. The engineering of the constructs used in the transformations, as well as the transformation procedure, are described in Tricoli *et al*. Based on greenhouse testing and small-scale field tests, two lead lines, ZW20 and CZW3, were chosen as parents of commercial hybrids. The characteristics of these lines are described in detail by Tricoli *et al*.

## ZW20

Line ZW20 was transformed with disarmed Agrobacterium harboring a binary Ti-plasmid. This plasmid contained expression cassettes consisting of the coat protein genes of zucchini yellow mosaic virus (ZYMV) and watermelon mosaic virus 2 (WMV2) under the control of the cauliflower mosaic virus 35S promoter. The T-DNA also contained, as a selectable marker, a neomycin phosphotransferase II (NPTII) gene from E. coli transposon Tn5, under the control of the cauliflower mosaic virus 35S promoter. The transformation event contained multiple insertion events, some of which were structurally complex. Among these insertion events were two loci containing genes encoding both the ZYMV and WMV2 coat proteins, but not the NPTII gene. One locus provided a moderate level of resistance to the two target viruses, while the other provided a high level of resistance. Advanced breeding lines were made homozygous for the high-resistance locus, but were segregating for the moderate-resistance locus. During this process, the NPTII marker was also eliminated by selection of lines that not only were homozygous for the high-resistance locus, but had also lost the selectable marker through segregation.

## CZW3

Line CZW3 was transformed with the plasmid pPRBN-CMV73/ZYMV72/WMBN22, containing expression cassettes for the coat protein genes for ZYMV, WMV2 and cucumber mosaic virus (CMV), as well as the NPTII selectable marker. Line CZW3 was a simpler transformant, consisting of only one insertion event, containing all three coat protein genes, as well as the selectable marker. Advanced breeding lines were homozygous for the transformed locus.

## Field Results

Field trials at a number of locations showed that the lines ZW20 and CZW3 were effective in controlling infections by ZYMV, WMV2, and - in the case of CZW3 - CMV as well. Tricoli *et al.* mechanically inoculated two different ZW20 sublines, including one homozygous for the high-resistance locus. This subline showed only 20% infection after inoculation with WMV2, and no infection at all after ZYMV infection, compared with 100% infection of non-transgenic controls.

The same researchers studied the resistance of CZW3 after mechanical inoculation with CMV, ZYMV and WMV2 individually, and in a mixed infection. Segregating inbred lines were tested, as well as a hybrid in which the segregating inbred was used as a parent. After individual and mixed inoculation of transgenic segregants, only one plant out of a total of 146 showed mild symptoms, while all of the 103 non-transgenic segregants as well as all of the 259 non-transgenic control plants became severely symptomatic.

Further tests were carried out in the field, comparing virus resistance in these lines after mechanical and aphid inoculation. Other tests studied the economic benefit of the resistance provided by lines ZW20 and CZW3 either as an inbred line or as the parent to a hybrid. These tests confirmed that virus resistance conferred by the coat protein genes was able to dramatically increase crop yield and economic return under the pressure of naturally vectored disease in the field.

## USDA EXEMPTION

The determination of efficacy against viral disease under field conditions led to the preparation of a petition for a determination of non-regulated status from the United States Department of Agriculture, Animal and Plant Health Inspection Service (USDA-APHIS). USDA-APHIS is the primary agency responsible for regulating work, in the greenhouse and field of transgenic plants. It also has the authority to grant non-regulated status to transgenic plants that in its opinion no longer present a plant pest risk.

The main responsibility for conducting an assessment of safety rests with USDA-APHIS. The assessment relies upon a review of the information package supplied to the agency by the petitioner, but the agency can and does seek out information on its own, and relies on the expertise of the reviewers on its staff. All relevant information is used by the agency to conduct an environmental assessment, and - if the assessment allows - grant non-regulated status.

One guiding principle followed by the USDA-APHIS in assessing risk is to compare the characteristics of the transgenic plant with the characteristics of plants that are produced by traditional plant breeding. Plants produced by traditional breeding are not regulated, because history provides sufficient assurance of safety. Thus, if the transgenic plant were determined to pose a risk different from that posed by plants already existing in nature or plants that would be produced by traditional breeding, then non-regulated status would be withheld. If, on the other hand, the risks presented by the transgenic plant are not different from those that are presented by traditionally bred varieties, then non-regulated status can be granted.

Since USDA-APHIS relies heavily (but not exclusively) on the information contained in the petition for non-regulated status, information in such a petition should include the following:

- A description of the biology of the plant prior to the transformation. The biological description should include detailed taxonomic information.
- Relevant experimental data and publications relating to the work conducted on the transgenic plant.
- A description of the differences in genotype between the transgenic plant and the corresponding non-transgenic plant. Information must be provided on the organisms from which the transgenes or other control elements were obtained, the method of transformation and the vectors used during the transformation process, as well as the inserted genetic material and its products. USDA-APHIS, the Canadian Food Inspection Agency, and Health Canada, have standardized the types of molecular and expression data submitted in a petition for non-regulated status and the criteria by which they should be judged.
- A description of the phenotype of the transgenic plant, including any known and potential differences from the corresponding non-transgenic plant, in order to substantiate the claim that the transgenic plant is unlikely to pose a greater plant pest risk than the non-transgenic plant from which it was derived.

This information should include the following: plant pest risk characteristics, disease and pest susceptibilities, expression of the gene product, new enzymes, or changes to plant metabolism, weediness of the transgenic plant, impact on the weediness of any other plant which is sexually compatible with the transgenic plant, agricultural or cultivation practices, effects of the transgenic plant on non-target organisms, indirect plant pest effects on other agricultural products, transfer of genetic information to organisms with which it cannot interbreed, and any other information which might be relevant. Any information indicating that a transgenic plant may pose a greater plant pest risk than the corresponding non-transgenic plant must be included. Field test reports for all trials of the transgenic line in question that were conducted under USDA-APHIS permit or notification. The field test reports should include any observations of deleterious effects on the environment caused by the transgenic plant. Based on information received by USDA-APHIS from various sources including the petition, the Agency granted non-regulated status to ZW20 and CZW3, and published their support for this decision.

In keeping with their goal of considering all information that might be available to them, USDA-APHIS made use not only of data supplied by the petitioners, but also relied on the literature, external expert opinion, and comparison with traits that were already familiar to agriculture. This approach acknowledged that scientifically valid risk assessment could be conducted by relying on information obtained from a wide variety of sources.

## Weediness

The particular concern in determining the plant pest risk of transgenic squash was the potential for the crop itself to become a weed, or for new weeds to arise as a consequence of gene flow from the crop to sexually compatible wild populations. This question was especially important because the transgenic squash would be released in a region that was known to be the centre of origin for C. pepo.

After evaluating all the information available, USDA-APHIS concluded that ZW20 and CZW3 were unlikely to cause a weed problem or a threat to genetic diversity of free-living squash if released from regulation. With regard to causing a weed problem, the agency concluded that the transgenic lines were unlikely to become weeds themselves, and that the virus-resistance traits were unlikely to cause free-living squash to become weeds.

The conclusion that the transgenic lines themselves were unlikely to become weeds was based on three reasons. First, the inbred lines used in the transformations belonged to the yellow crookneck squash type (C. pepo ssp. ovifera var. ovifera), which did not exhibit weedy characteristics and was not considered to be a weed in standard texts and weed lists.

Second, introduction of CMV (for CZW3) as well as ZYMV and WMV2 resistance into the crop was not different from other disease- or pest-resistance traits already present in the C. pepo gene pool or introduced into it by traditional breeding. These traits had already been introduced into squash, and even into some commercial cultivars, yet these cultivars had not become weeds. Finally, the two transgenic lines did not exhibit any other characteristics that were more 'weedy' than those exhibited by traditionally bred squash cultivars. Aside from considering the potential problem caused by the transgenic lines themselves, USDA-APHIS also evaluated the possibility that traits transferred from ZW20 and CZW3 to free-living relatives would cause those free-living plants to become more 'weedy'. USDA-APHIS concluded that this problem was unlikely to happen. The agency assumed that the transgenic lines and their cultivated progeny would be able to pollinate free-living squash plants and transmit the virus resistance traits to them. This undoubtedly had occurred in the past with other disease- or pest-resistance traits that had been introduced into C. pepo. However, despite this opportunity, no weedy hybrid progeny had emerged.

Since there was no evidence to suggest that the transgenic traits would behave differently from the traditionally bred traits, the historical lack of weed problems provided a sufficient level of assurance that crosses between transgenic and free-living squash were unlikely to create a weed problem as well. Furthermore, the cultivated squash genotypes that were used for transformation possessed characteristics that placed those plants at a selective disadvantage in a non-cultivated habitat. The transmission of these traits to free-living relatives along with the transgene would place recipient plants at a disadvantage, further reducing their weedy potential.

Finally, surveys of free-living squash populations taken by the applicant failed to detect infections of CMV, WMV2 or ZYMV. Therefore, there was no evidence that viruses limited the populations of free-living squash. Consequently, it was not likely that virus resistance would release these populations from control, nor was it likely that virus resistance conferred a significant selective advantage in free-living habitats. The lack of selective pressure exerted by viruses was also cited by USDA-APHIS as the reason for concluding that genetic diversity in free-living populations would not be reduced. Without selection pressure, there was no reason to expect selection of a subset of genotypes into which the virus resistance would have been crossed.

## TRANSCAPSIDATION/RECOMBINATION

The commercial introduction of viral sequences via transgenic plants - and in particular viral genes encoding viral coat proteins - present the potential for interactions between the transgenes or their products, and the viruses to which the transgenic squash might remain susceptible. These interactions include transencapsidation of heterologous viral RNA by coat proteins expressed in the plants, phenotypic mixing between viral coat protein subunits expressed by the transgenic plant or by the transgenic plant and infecting viruses, and recombination between transgenic viral components and other viruses infecting the transgenic plants.

DeZoeten pointed out that these phenomena have the potential to produce new viruses. USDA-APHIS concluded that the risk of transencapsidation and phenotypic mixing in transgenic virus-resistant plants would be no greater than their likelihood of occurrence in non-transgenic plants already in agriculture. Data submitted by the petitioners, and other information gathered by the agency, showed that the amounts of viral coat protein produced in transgenic plants resistant to CMV, ZYMV and WMV2 were less than the amounts found in susceptible, virus-infected plants.

With respect to this set of data, USDA-APHIS noted that ZYMV and WMV2 coat protein levels increase when ZW20 plants were inoculated with ZYMV and WMV2 singly or combined. The agency speculated that this increase might be due to limited replication of the viruses in inoculated ZW20 plants, or because the coat proteins produced by the transgenes were stabilized by limited replication of the inoculated viruses.

## EPA EXEMPTION FROM TOLERANCE

As is the case with USDA-APHIS, the US Environmental Protection Agency (EPA) also has the primary responsibility of conducting a safety assessment, based on information supplied by the developer of a transgenic crop and other information that the agency might obtain, including that of external and internal experts. When the first transgenic squash line (ZW20) was ready for commercialization, the EPA did not have a final rule covering transgenic plants. However, consultation with the EPA took place during the time the agency was formulating its proposed rules.

These rules extended the agency's jurisdiction to all plants expressing disease-or pest-resistance traits, since they could be viewed as producing a substance that was pesticidal - that is, they mitigated the effect of a pest. While the agency claimed jurisdiction over these plants, it recognized that certain broad categories of plants could be exempted from regulation.

For example, plants produced through traditional breeding, or plants that repelled pests or disease through the modification of some structural feature could be exempted since they usually presented no new pesticidal exposure to humans consuming those plants. Furthermore, the EPA proposed to exempt coat protein produced in transgenic plants from regulation under the Federal Insecticide, Fungicide, and Rodenticide Act (FIFRA) and the Federal Food, Drug, and Cosmetic Act (FFDCA), again because they presented no new exposures to humans consuming them. Even though the rules were only proposed, consultation with the EPA led to the decision to submit a petition for exemption from tolerance (the legal limit for a pesticide chemical residue in or on a food), as required by the FFDCA prior to the commercialization of any pesticide. In order to obtain this exemption, it was important to show that the establishment of a tolerance level was not required to protect the public health.

This goal was achieved for line ZW20 by measuring the amount of viral coat protein in samples of various melon and squash fruit on sale at major grocery stores in one representative locality in the United States. Results showed that levels of viral coat protein in transgenic fruit were within the range found for most of the fruit in the sample. For some samples, the fruit on the market had many-fold higher levels of coat protein than in the transgenic squash.

These data were used to support the claim that viral coat proteins in general, and ZYMV and WMV2 coat proteins in particular, were not substances that fitted the two main criteria which the EPA planned to use to determine whether a pesticidal substance would require review. First, in order to trigger EPA review the pesticidal substance would be one that was not derived from a known food source. Second, the substance would present the following changes in pesticidal exposure to humans: it would be found in significantly higher levels in the transgenic plant than that normally found in food, it was being transferred to a different crop than that in which it was already found, it was not already found in edible tissue, and it was not already found in unprocessed food.

Other arguments advanced in the petition in support of the exemption from the requirement of a tolerance were based on the premise that a demonstrated history of safe use was sufficient to assess safety. Examples from the literature showed that humans were already exposed to viral coat proteins in general, not only through consumption of naturally infected produce, but also through ingestion of food obtained from crops that had been subjected to cross protection (the intentional inoculation of a crop with a mild strain of a virus in order to protect the crop from subsequent infection by a more virulent strain of the virus). The petition also argued that supported the claim that expression of ZYMV and WMV2 in ZW20 would not present any changed exposure of humans to allergens. Because ZYMV and WMV2 coat proteins are commonly present in cucurbit fruit, anyone who might be allergic to viral coat protein in squash (a phenomenon for which no documented cases were known) would have the same reaction to the transgenic fruit as to non-transgenic fruit they had already consumed. The petition also addressed exposures due to dermal contact and inhalation. The petition argued that a history of safety had been established through the lack of reports of toxicity or allergenicity from researchers who had undoubtedly been exposed to these viruses by both the dermal and inhalation routes. Finally, the petition pointed out that viral coat proteins were ubiquitous in the environment, and therefore a natural component of it.

Infected cucurbit fruit that are not marketable are routinely plowed back into the soil, thus adding large amounts of viral coat protein to the environment. Furthermore, studies in the literature have reported the recovery of plant viruses from ground water runoff in Europe and North America. The EPA considered the data and the arguments presented in the petition and granted an exemption from tolerance for the ZYMV and WMV2 coat proteins produced within ZW20 plants. Subsequently, the EPA granted a more categorical exemption from tolerance for ZYMV and WMV2 coat proteins in all food commodities. The decisions and supporting reasoning are summarized in the following paragraphs, which are based on the published decision. A detailed review of this decision is especially instructive because it provides insight into the issues that by law must now be considered by the EPA.

A detailed review also provides another example of the reliance of US regulatory agencies not only on data submitted by petitioners, but also on the literature and other sources of information. In granting the broader exemption from tolerance, the EPA considered several issues that by law they were obligated to address, especially as a result of the passage of the Food Quality Protection Act of 1996 (FPQA), which amended the FFDCA. The amended FFDCA allowed EPA to establish an exemption from the requirement for a tolerance only if EPA determines that the tolerance is 'safe'.

By law, 'safe' meant that there was 'a reasonable certainty that no harm will result from aggregate exposure to the pesticide chemical residue, including all anticipated dietary exposures and all other exposures for which there is reliable information'. This includes exposure through drinking water and in residential settings, but does not include exposure due to its use during application or other occupational activities. In deciding on exemptions from tolerance the EPA is also required to pay special attention to exposure of infants and children. The EPA examined two areas of concern in determining the risks from aggregate exposure to pesticide residues. First, the agency examined the toxicity of the 'pesticides' (ZYMV and WMV2 coat proteins). Second, it examined exposure to the pesticide through food, drinking water, and through other exposures that occur as a result of pesticide use in residential settings. In considering toxicity, petitioners are required by law to submit data regarding acute toxicity, genotoxicity, reproductive and developmental toxicity, as well as subchronic and chronic toxicity.

However the petitioner requested, and EPA granted, waivers from these data requirements. The waivers were based on the long history of mammalian consumption of the entire plant virus particle in foods, without causing any deleterious human health effects. The EPA once again acknowledged that virus-infected plants currently are - and have always been - a part of both the human and domestic animal food supply, and that there is no evidence that plant viruses are toxic to humans and other vertebrates. Furthermore, the EPA pointed out that plant viruses are unable to replicate in mammals or other vertebrates, thereby eliminating the possibility of human infection. As additional assurance of safety, only the coat protein would be expressed in the plant, and this component alone would be incapable of forming infectious particles.

The EPA also had to consider the toxicity of the introduced DNA (coding and regulatory regions) directing the synthesis of the coat proteins. The EPA acknowledged that DNA is common to all forms of plant and animal life and that it was consumed as a safe component of food. The specific nucleic acids that were used to transform the squash plants were characterized by the petitioners, and a review of the nucleic acid sequence led EPA to conclude that no mammalian toxicity was anticipated from the genes encoding the coat proteins of ZYMV and WMV2. In considering exposures, the EPA had to consider available information concerning exposures from the coat protein in food and all other non-occupational exposures, including drinking water from groundwater or surface water, and exposure through its use in gardens, lawns or buildings. With respect to dietary exposure in food, the EPA concluded, as they did before, that the use of viral coat protein in squash would not result in any new dietary exposure to plant viruses, because:

- Entire infectious particles of ZYMV and WMV2 are found in the fruit, leaves and stems of most plants,
- Viruses are ubiquitous in the agricultural environment at levels higher than will be present in transgenic plants,
- Virus-infected food plants have historically been a part of the

human (including that of infants and children) and domestic animal food supply, with no known adverse effects to human or animal health.

Moreover, the EPA concluded that there was no evidence to indicate that harmful effects occurred as a result of exposure to viral coat proteins through other means, such as non-food oral, dermal and inhalation routes. They arrived at this conclusion by relying not only on data submitted by the petitioners, but on the expert opinion of its Scientific Advisory Panel, which concluded that the levels of virus in the agricultural environment are much higher than those present in transgenic plants, and that the existing presence of viral coat proteins in the food supply provided a scientific basis for exempting transgenic plants that express viral coat proteins from the requirement of a tolerance.

With respect to drinking water exposure, the EPA concluded that potential exposures through this route would be negligible. The agency reasoned that viral coat proteins produced in plants are an integral part of the living tissue of the plant, and are therefore subject to rapid degradation and decay. Consequently, viral coat proteins would not accumulate in the environment or in the food chain.

Therefore, because of their presumed transience, these proteins would present only negligible exposure in drinking water obtained from surface or groundwater sources. The EPA used similar reasoning to conclude that exposure of engineered coat proteins as a consequence of their use in such areas as around homes, parks, recreation areas, athletic fields and golf courses, would be negligible. The EPA concluded that the likelihood of exposure to plant viral coat proteins, including ZYMV and WMV2 coat proteins, through any dermal or respiratory routes would be unlikely.

Although physical contact with the plant or raw agricultural food obtained from the plant may present some opportunity for dermal exposure, the potential amounts involved in this exposure were judged to be negligible compared with exposure through the diet. Furthermore, the EPA reasoned that the skin would provide a barrier that would not be crossed by viral coat proteins. Similarly, exposure through inhalation was judged to be negligible compared with potential exposure through the diet.

The most likely opportunity for respiratory exposure through inhalation was through pollen. However, the EPA concluded that it was again unlikely that the amount of exposure to the coat proteins in the pollen would be significant. Viral coat proteins, when present in pollen, would be part of the pollen grain and would not cross the barrier presented by the mucous membrane of the respiratory tract. Therefore, this route would not add any additional exposure to that presented by the dietary route. A new requirement imposed by the Food Quality Protection Act involved the assessment of available information, indicating that there might be special concerns regarding exposure of infants and children to viral coat proteins.

The EPA concluded that viral coat proteins were present in all foods, including those foods consumed by infants and children. It had no evidence to indicate that the amount of exposure was different between adults and children, nor that the effect of viral coat proteins differed between the two groups. Most importantly, the EPA had no evidence suggesting that exposure of adults or infants and children was in any way harmful.

Finally, the EPA had to consider whether ZYMV and WMV2 coat proteins, or the DNA needed to produce them in plants, would have an effect on the immune and endocrine systems. Because the agency had no information indicating that such an effect could be expected, it did not require that the petitioner submit information regarding that question. The decision to grant an exemption from tolerance for ZYMV and WMV2 coat proteins in ZW20 and subsequently in all raw commodities also cleared the way for commercialization of CZW3.

EPA then granted an exemption from tolerance for the coat protein of CMV in all raw commodities, based on the same safety assessment it conducted for the other two coat proteins. Line CZW was included in this exemption and was therefore cleared for commercialization.

## FDA FOOD SAFETY ANALYSIS

In contrast to USDA-APHIS and EPA, the United States Food and Drug Administration (FDA) makes the safety assessment of new plant varieties, and the foods made from them, the responsibility of the developer or manufacturer. In 1992, the FDA published guidelines to clarify the obligations of any producer of new plant varieties, whether produced through traditional plant breeding, or through techniques of recombinant DNA.

The principle governing the guidelines is that assessment of the safety of a product (in this case food produced from a new plant variety) should focus on the product's characteristics rather than the method by which it was made. The authority of FDA to regulate food is a post-market authority. Thus, a food producer can market a food containing a new component that it has assessed to be generally recognized as safe (GRAS). If this component is not GRAS, then a food additive petition must be filed and granted before food containing that new component can be marketed. The 1992 guidelines were meant to help producers in determining whether a food safety concern existed. If a concern existed, a consultation with FDA would take place to determine whether a new component could be considered GRAS, or whether a food additive petition was required. In practice, consultation with FDA regarding new transgenic plants that will become food usually occurred regardless of safety concerns.

The consultation assures producers that FDA is aware of the developer's safety assessment process, and serves to identify those parts of the safety analysis that might not be acceptable to the Agency. This process enables the producer to proceed toward commercialization with confidence that no unanticipated safety concerns will be encountered at late stages of product development. Recently, the FDA has published its intention to make this consultation process mandatory.

The assessment applies equally well to CZW3. In general, the safety assessment specified by the guidelines examines the host plant (the plant which has been modified by introduction of foreign DNA), the organisms from which the foreign DNA has been obtained, and the new compounds produced as a result of the introduction of the foreign DNA. First, the guidelines direct the developer to determine whether there is a history of safe consumption of the species that was transformed.

In the case of squash, the species has had a history of safe consumption for at least 3000 years, and is consumed in many forms throughout the world. The particular genotypes transformed to produce ZW20 and CZW3 have been in use commercially for several years. The guidelines also ask the developer to consider whether characteristics of the host species, related species, or the lines used in transformation warrant analytical or toxicological tests. If so, they also ask whether the results of the tests provide evidence that the toxicant levels in the new plant variety do not present a safety concern.

Squash indeed has characteristics warranting analytical testing, since the plant is known to produce cucurbitacins, which are extremely bitter toxic alkaloids. Squash breeders are aware that cucurbitacins must be eliminated from commercial varieties. Therefore, standard breeding practices take into consideration the potential presence of this compound, especially if genes from wild or free-living relatives are crossed into cultivated squash.

The FDA guidelines indicate that the types of testing currently used to screen for known toxicants have been effective in managing the risk of such compounds, and are therefore appropriate to apply to the case of transgenic crops. For the cucurbitacins, tasting fruits for bitterness has been an effective test method for the detection of this compound. The documented sensitivity of the human sense of taste for this compound is as low as 1-10 parts per billion, depending upon the particular form of cucurbitacin. At 10 parts per billion, this detection level is 34 000 times less than the reported oral LD50 in mice. Therefore, tasting fruit is an appropriate analytical test for the presence of cucurbitacin, and the presence of these compounds can be detected long before toxic levels are reached.

The non-bitter taste of ZW20 and CZW3 indicates the absence of cucurbitacins in these lines, as they were in the original lines before transformation. This result CMV, ZYMV and WMV2 coat proteins are almost identical to the coat proteins (presumably edible) found in cucurbit fruit. Squash containing viruses is not known to be allergenic, nor is it known to be toxic. Intake of viral coat proteins from the fruit of transgenic plants will be comparable to the intake of these proteins in non-transgenic infected fruit.

Finally, the amounts of viral coat protein expressed in transgenic ZW20 and CZW3 are too low to present a possibility of becoming a major dietary constituent. These data were presented to the FDA in order to inform the Agency of the justification for the judgement that no food additive petition was warranted for ZW20 and CZW3 (they were therefore GRAS). Since no objections to these analyses were raised by the Agency, commercialization proceeded.

The assessment of safety of transgenic crops continues to be a topic of scientific and public concern. A review of the issues that surround the debate over the safety of these crops reveals that the issues addressed seven years ago during the commercialization of virus-resistant squash - which was the first commercialized virus-resistant crop - are still unresolved. The appropriate perspective on safety is one that was expressed well before transgenic crops were at issue: A thing is safe if its risks are judged to be acceptable. By its preciseness and connotative power, this definition contrasts sharply with simplistic dictionary definitions that have 'safe' meaning something like 'free from risk.' Nothing can be absolutely free of risk.

Because nothing can be absolutely free of risk, nothing can be said to be absolutely safe. There are degrees of risk, and consequently there are degrees of safety. This is the perspective that is taken by the agencies responsible for regulating transgenic crops in the US. They all judge the level of risk of these crops by comparing them to the existing level of risk, such as risks presented by crops developed by traditional breeding or risks presented by natural phenomena.

Furthermore, even though the risk presented by these crops might be greater than those presented by current unregulated activities, the acceptability of that risk also determine whether these crops are safe enough to commercialize and use. It is this latter issue that adds the dimension of public perception and acceptance to the debate.

The deployment of transgenic crops will most likely continue and increase with time. The potential benefits to the consumer, the farmer, as well as to the companies developing these crops will serve as a potent force for the exploitation of this technology. As this continues, scientific inquiry into the potential risks should continue. These investigations will help to further clarify areas of real concern. By focusing attention on these areas, which will probably be very few, safe use of this technology will be further improved.

## RESISTANCE TO PLANT VIRUS DISEASES

### PROTEIN

One successful application of genetic engineering in crop improvement is in the production of virus-resistant plants. This has been achieved by genetic transformation of plants with novel resistance genes based on nucleotide sequences derived from the viruses themselves or from virus-associated nucleic acid as well as genes from other sources. The demonstration that the expression of tobacco mosaic virus (TMV) coat protein (CP) in transgenic plants protects the plants against virus infection has led to explorations of using other viral or virus-associated genes in producing genetically engineered resistance against in plant virus. Transgenic plants expressing Sat-RNAs have been obtained in a number of labouratories and have been tested in greenhouses and fields. Recent studies with defective interfering RNA or DNA protection and sense-antisense-ribozyme RNAmediated protection offer much promise for a broad range virus resistance in plants. Our labouratory has successfully produced satellite RNA-mediated resistance to cucumber mosaic virus (CMV) in tobacco and tomato, CP-mediated resistance to rice stripe virus in rice and fusion protein (CP and nuclease) mediated high resistance to tobacco mosaic virus in tobacco. One way to confer more effective and durable field resistance to virus disease in transgenic crop is to use a combination of multiple resistance genes.

### ENGINEERING RESISTANCE TO PLANT-VIRUS DISEASES

Since 1981, Sat-RNAs have been used as a biological control agent (BCA) of diseases caused by CMV on a large scale in China. The results show that the BCA is effective in controlling diseases caused by virulent CMV strains in many crops and, in addition, induces resistance to certain types of fungal diseases. The protective effects of CMV Sat-RNAs have been demonstrated in a number of greenhouse and field tests. These studies have provided the basis for engineering plant resistance to virus infection by introducing the Sat-RNA as a transgene. In our labouratory, the cDNA of an attenuating Sat-RNA monomer from CMV strain 1 was synthesized and constructed into the plant expression vector pRok2. The Sat-RNA cDNA was introduced into different plant species including tobacco and tomato.

When transgenic tobacco and tomato plants were tested in the greenhouse for resistance against challenge infection by CMV, they showed a significant decrease in disease index compared to nontransformed plants. In an experimental scale field test conducted in the spring of 1990, the transgenic tomato plants inoculated artificially with CMV gave about 50% higher fruit yield than the non-transformed control plants and showed a decrease in disease index.

However, there was no significant difference in leaf yield between the satellite transgenic and the control tobacco plants, although the disease index of transgenic tobacco was about 10% lower than that of the control. Due to an insufficient level of Sat-RNA in the plants at an early stage of infection, the transgenic tobacco plants showed only a weak resistance to CMV, which directly affected the leaf yield, since it is the early leaves that are harvested in tobacco production. Although Sat-RNAs confer some extent of resistance to virus infection, field tests carried out in China from 1990-1992 show that the resistance of the transgenic plants is not strong enough to fully protect tobacco and tomato plants from the damage caused by natural virus infections.

## SAT-RNA IN A MULTIPLE GENE STRATEGY OF ENGINEERING RESISTANCE

One way to obtain more effective and durable resistance to virus diseases in transgenic plants may be to use a combination of multiple resistance genes. In this strategy, different genes that produce blockage or interference at different stages of a virus infection are used to obtain engineered resistance.

Transgenic lines of tobacco expressing both CP and Sat-RNA of CMV were obtained through transformation using a chimeric vector containing two expression cassettes of the two genes under the control of the CaMV 35S promoter. When challenged with CMV, the virus concentration of challenging CMV in transformed plants that express only Sat-RNA was about 10-20% of that in non-transformed plants, while in CP + Sat-RNA transformed plants it was only 4-5%. A comparison of disease incidence and disease index of CP + Sat-RNA, CP alone or Sat-RNA alone transformed tobacco plants. The resistance of CP + Sat-RNA transformed plants was about twice as strong as that of Sat-RNA alone or CP alone transformed plants. Since the CP level expressed in CP + Sat-RNA transformed plants was similar to that in transformed plants that contained only the CP gene, the two-fold increase in resistance of CP + Sat-RNA transformed plants was not due to higher levels of CP expression.

On the other hand, the level of Sat-RNA accumulation in the CP + Sat-RNA transformed plants was lower than that in the Sat-RNA alone transformed plants. This may be due to the effect of the CP gene conferring resistance at an early stage in CMV infection, reducing the replication of CMV as well as Sat-RNA. Preliminary data from a field test of 1992 confirmed the enhanced resistance of the CP + Sat-RNA transformed plants to CMV. The CP + Sat-RNA transformed plants show stronger resistance to natural CMV infections than either the CP alone or Sat-RNA alone transformed plants.

## TRANSGENIC TOBACCO PLANTS RESISTANT TO CUCUMBER MOSAIC VIRUS

Since Nicotiana tabacuum is an amphidipoid species, the tetraplold transgenic plants will segregate extensively in their progenies. It needs to be propagated several generations before obtaining a homozygous line. In order to speed up this process, a procedure for a fast production of homozygosity of transgenic plants was devised in our work. Leaf discs of diploid tobacco plant derived from anther cultures of a commercial cultivar NC-89 were transformed with a chimeric vector of CP and Sat-RNA genes of CMV.

## RESISTANCE TO RICE STRIPE VIRUS IN TRANSGENIC INDICA RICE PLANTS

Rice stripe disease occurs in rice growing areas of China, Japan, Korea and the Commonwealth of Independent State, and causes significant reduction in rice yield. The coat protein (CP) gene of a Chinese isolate of rice stripe virus (RSV-C) was synthesized, cloned and sequenced, which is very similar to the Japanese strain. Cell suspension cultures were initiated from embrogenic calli of rice Annong S-1 (indica rice) by inoculating yellowish, compact and embryogenic calli derived from seeds into a suspension culture medium containing proline and maltose. After being cultured at 26°C in the dark for about half a year, finely-dispersed and embryogenic suspension culture was established. Before bombardment, the suspension culture was evenly applied onto three-layerfilter-paper discs in a petri dish. Cells were bombarded with 1 ìm diameter tungsten particles coated with DNA of the expression vector pROK2 containing the CP gene under control of the 35S promoter.

The plasmid DNA was absorbed to tungsten particles using a calcium-spermidine precipitation procedure. 2.5 ìL of the coated particles was loaded onto macroprojectile and each dish with the suspension culture was bombarded three times under a partial vacuum. Following bombardment, the suspensions were cultured in a modified N6 medium. Two days later the suspensions were transferred to the same medium containing G418 (40 ìg/mL), which were subcultured every week.

Being subject to G418 selection for two months, white and fast-growing colonies were emerged from the browny cultures. Green plants could be regenerated when the resistant calli were transferred to a differentiation medium. In one experiment ten plantlets regenerated from G418 resistant calli were tested for their transgenic nature by Southern blot analysis using a32P-dCTP-labelled CP gene as a probe. The genomic DNA extracted from the selected as well as the control plants were digested with EcoR1 and BamH1 and then hybridized with the probe.

Two plants showed two hybridization bands of 0.6 kbp and 0.7 kbp corresponding to the CP gene. Western blot and ELISA further analyses demonstrated that CP (32 kDa) was expressed in the transgenic rice plants. Sixteen transgenic thus obtained and 100 non-transformed rice plants were inoculated with RSV-carrying leaf hoppers, 24 days post inoculation feeding, most of the untransformed plants developed chlorotic stripes and chlorosis in the young leaves, the disease index (DI) being 53%.

After 40 days post inoculation feeding, most of the control plants showed severe symptoms, being stunted and the leaves developing chlorotic and brown necrotic streaks, with the DI being 88%. In contrast, the transgenic rice plants subject to the inoculation feeding developed no or mild symptoms. After 24 and 40 days post inoculation feeding, the disease index of the transgenic plants was 15% and 25% respectively.

## INCREASE OF RESISTANCE

Coat protein-mediated resistance (CPMR) has been proved to be successful against several families of plant viruses. Recently CPMR was obtained by expressing a truncated CP of tobacco virus. Our work has indicated that expression of a fusion protein of TMV-CP and Staphylococcal nuclease (SN) in transgenic tobacco plants significantly enhances the CPMR to TMV. TMV-CP or SN gene alone and CP'!SN or SN'!CP fusion protein genes were obtained by PCR, inserted downstream of 35S promoter of a binary vector pROK2 and transformed into tobacco leaf discs.

No SN gene alone transformed tissues could be regenerated, but the tissues containing CP'!SN or SN'!CP fusion genes grew normally and could be regenerated into plants. The transformed nature of the regenerated tobacco plants were confirmed by Southern blot hybridization and PCR analysis.

It was demonstrated that the fusion proteins of CP'!SN and SN'!CP were expressed in the transgenic plants by Western blot analysis. CP'!SN- or SN'!CP-transformed plants exhibited higher resistance to TMV than CP alone transformed ones. 50% of CP'!SN- and SN'!CP-transformed tobacco plants were symptomless for three months after inoculation with 1 ìg/mL TMV.

Nuclease activity test of the fusion protein expressed in transgenic plants was carried out in vitro. Fusion proteins of SN and CP were extracted from the transgenic plants and were obtained by immunoprecipitation with antiserum of TMV.

The precipitated proteins were infiltrated with 2% SDS and then with water containing excess amount of TMV-CP, and reacted at 30°C with the incubation mixture of 25 mM Tris-HCl (pH 8.8), 10 mM $CaCl_2$ and 40 ng 32P-5'-end-labelled oligdeoxynucleotide. The result showed that no nuclease activity was observed. No obvious difference in structure and infectivity of virion isolated from SN-CP transgenic and non-transgenic plants has been found.

The mechanism of resistance to virus by expression of a fusion CP was unknown. We speculate that the fusion CP is somehow dysfunctional and more effective to disrupt the normal virus-host relationship than CP does alone. The disruption may be at the level of virus movement throughout the plant or the fusion CP is encapsidated into virion to generate defective virus particle.

## RISKS ASSESSMENT OF RELEASING GENETICALLY ENGINEERED PLANTS

Some transgenic organisms, such as micro-organisms and animals will be grown under controlled conditions and only released to the environment accidentally. However transgenic plants will be grown on a large scale in the open environment. There is concern that transgenic plants may produce risks.

## Spread of Transgene by Pollen

Because genetic engineering will often result in the introduction of genes not previously present in a species, there is concern that such genes may spread to other plants and to weeds by crosspollination. It is important to have information on pollen transfer distances and the range of species that can be pollinated. To obviate this problem trial release of transgenic plants have often required the removal of flowers to prevent pollen spread as for the transgenic tobacco production. However, if such plants are to be used for the seeds or fruits, unlimited spread of pollen will occur.

## Releasing of Antibiotic Resistance Genes

Antibiotic resistance genes are often used as a selective marker in transformation procedure. In the case of antibiotic resistance, it will be a problem only if it is inherited by pathogenic micro-organisms which are being controlled by the antibiotic in question. The risk of a pathogen inheriting and being able to express an antibiotic resistance gene derived from a plant is at least a million-fold lower than such resistance being obtained from other micro-organisms.

## Pollution of Honey

A further problem that might be encountered is the presence of cloned gene products in pollen. This could harm the bee larvae which are fed pollen, pollutes honey with minute amounts of the gene products, and makes the pollen more allergenic. Most gene products will not be found in or on pollen in other than minute amounts and thus contamination should not be a problem.

## Transgenic Plants Becoming Weedy

Potentially the greatest risk is that certain types of transgenic plants might become weedy, for example, the acquisition of herbicide resistance genes which will make them less easy to control in subsequent generations. On the one hand herbicide resistance genes are used as selective markers in many labouratories; on the other hand large companies are attempting to encourage growers to use more herbicides, and are using genetic engineering for purely commercial reasons.

## Safety of Application Sat-RNA for Virus Diseases Control

Safety concerns about the use of Sat-RNAs remain. Even though virus replication is inhibited in transgenic plant, a small amount of Sat-RNA can still be encapsidated into helper virus particles and transmitted to other plant species, which it may cause more severe diseases. Moreover, because the exacerbating and attenuating Sat-RNAs can differ in only a few nucleotides, there is always a possibility that a symptom attenuating Sat-RNA can mutate into an exacerbating one. During a decade of using Sat-RNAs as biological control agents in China, a number of facts concerning the safety of using Sat-RNA in agriculture have been realised. In spite of the large scale application of the BCAs on different crops, no necrogenic or other harmful Sat-RNA variant was found in a systematic field survey in which random plant tissue samples were analysed by temperature-gradient gel electrophoresis (TGGE) to detect Sat-RNA sequence variants.

Secondly, the efficiency of aphid transmission of CMV in the presence of the Sat-RNA is reduced to about 10% of that in the absence of the Sat-RNA. The reduced transmission efficiency, due to the low concentration of CMV when the Sat-RNA is present, lowers the chance that the Sat-RNA will be released in the field. Finally, it is possible to maintain the benign property of the Sat-RNA by periodic and detailed quality control of the BCA stocks. It is our hope that the use of Sat-RNA as a viral control strategy will become increasingly safe as our knowledge of Sat-RNA structurefunction relationship increases. Indeed, it has recently been found that the Sat-RNA of CMV can lose its ability of being encapsidated into CMV CP by changing a few nucleotides; this in turn excludes the possibility of Sat-RNA transmission (D. Baulcombe, personal communication).

## Safety of Viral Coat Protein-mediated Resistance

Transencapsidation of viral nucleic acid by transgenically expressed CP may induce heteroencapsidated virion, which might be transmitted by new vectors and to new host plants. The CP of plum pox potyvirus has been shown to confer aphid-transmissibility on a non-transmissible isolate of zucchini yellow mosaic virus. However, recent studies showed that intentionally truncated, antisense or nonexpression (-AUG) CP genes could provide measurable protection or even complete immunity against the appropriate parent virus; this in turn excludes the possibility of heteroencapsidation of viral nucleic acid by transgenically expressed CP. In conclusion, safety, from a public view, was ensured by biologically containing the transgenic plants so that they would not induce risk.

The emphasis on biological containment has meant that for release to be permissible it is necessary to demonstrate beyond reasonable doubt that the transgenic plants are safe. But nothing is completely safe, and safety per se is not always desired; for example, plant species without any toxic metabolites would be exquisitely sensitive to pests.

## Chapter 5

# Canola Tolerant to Herbicides

Glyphosate (N-phosphonomethylglycine), the active ingredient in Roundup herbicide, is a broad-spectrum, post-emergent herbicide that is highly regarded for its efficacy and environmental safety. Until recently, Roundup has not been available for in-crop, selective control of the weeds which affect the yield and quality of canola crops. Developments in biotechnology have made it possible to insert two genes into canola which provide tolerance to glyphosate and thus offer Roundup Ready Canola (RRC) to Canadian farmers.

Before RRC could be used by farmers, its food, feed and environmental safety had to be established and approved by the Canadian regulatory authorities. The focus of the safety assessment was to establish that RRC was 'substantially equivalent' to conventional canola, allowing for the tolerance to glyphosate.

This paper describes the results of compositional studies that were conducted with RRC canola grown in 1992 and 1993 in Canada. The compositional studies measured the levels of important nutrients in canola seed, processed canola meal and in canola oil.

The levels of anti-nutrients in canola seed and in processed meal were also determined. Results of a four-week rat feeding study with RRC canola processed meal that confirm 'substantial equivalence' are also presented. The original line (transformation event) upon which the safety tests were performed is termed GT73. It is a selection from Brassica napus L. cv. Westar transformed using Agrobacterium tumefaciens. Line GT73 expresses two proteins conferring tolerance to Roundup herbicide.

The proteins are CP4 5-enolpyruvylshikimate-3-phosphate synthase (CP4 EPSPS) and glyphosate oxidoreductase (GOX) and they confer tolerance through two different mechanisms. The first tolerance mechanism involves a decreased sensitivity of the site of herbicidal action to glyphosate accomplished through the discovery of CP4 EPSPS, an EPSPS with a significantly higher Ki for glyphosate. The presence of CP4 EPSPS protein in RRC enables the plant to make aromatic amino acids despite application of glyphosate, which inhibits endogenous EPSPS. The second mechanism entails the breakdown of glyphosate to aminomethylphosphonic acid (AMPA) and glyoxylic acid by GOX. Line GT73 has demonstrated commercial levels of tolerance to Roundup herbicide in numerous field trials. Roundup herbicide has a long history of safe use in agriculture based on the low toxicity of its active ingredient, glyphosate, to a wide variety of vertebrate and invertebrate species. AMPA has a safety profile similar to that of glyphosate.

*The safety assessment of the CP4 EPSPS enzyme introduced into canola has been addressed elsewhere. CP4 EPSPS protein was shown to be*:

- Readily degraded in simulated digestive fluids;
- Non-toxic when administered orally to mice at an acute dose thousands of times higher than potential human exposure to CP4 EPSPS in foods;
- Not structurally or functionally related to known protein allergens or toxins based on amino acid sequence homology searches of protein toxin and allergen databases.

GOX enzyme has been subjected to the same safety evaluation and has also been shown to be safe for consumption in food or feed (data not presented). Since the use of biotechnology to develop new crop varieties is relatively new, nations such as Canada have developed guidelines to assess the safety of genetically modified plants.

These guidelines are science-based, applied case-by-case, and are designed to assess the feed and environmental safety of the plant (regulated by the Canadian Food Inspection Agency) as well as the safety of human food derived from the plant (regulated by Health Canada). Like most national guidelines, the Canadian guidelines are founded on the concept of 'substantial equivalence'. The premise of 'substantial equivalence', as defined by the Organisation for Economic Co-operation and Development and the World Health Organization and used in Canada to regulate genetically modified plants, is that existing plant varieties can be used as a basis of comparison to establish safety. It is presumed that if the agronomic performance and nutrient/ anti-nutrient composition of GT73 is similar to that of B. napus canola from which it is derived, GT73 would be considered substantially equivalent and 'as safe as' other canola varieties.

This assumes that the safety of proteins introduced into GT73 have already been established. Furthermore, new B. napus varieties of RRC that have been developed using traditional breeding and have not been otherwise modified will be regulated as any other canola variety. We assessed the substantial equivalence of GT73 in field tests evaluating its agronomic performance (data not presented) and extensive compositional analyses of key nutritional and anti-nutritional components known to exist in canola. Agronomic performance was assessed in Western Canadian Canola/Rapeseed Recommending Committee Co-Op Tests conducted in 1993 at 24 test locations; test plots were replicated four times. This paper focuses upon summarizing the compositional analyses in addition to confirmatory feeding studies conducted with rats. Substantial equivalence was concluded when results obtained for GT73 were within commercially acceptable ranges for B. napus varieties of canola.The principal comparison was to the limits that have been established for canola and to the ranges published for B. napus. Data for Westar from Monsanto-conducted field trials were used to assess substantial equivalence. Additional data for Westar from the 1992 and 1993 Western Canadian Canola/Rapeseed Recommending Committee Co-Operative Test (Co-Op Test) was also used extensively for comparisons. Results of a four-week rat feeding study with an RRC line (RU3) derived from GT73 canola are also presented. In this study, results from a feeding of RU3 to rats are compared to those from feeding the parental (non-transformed line) and commercial canola/oilseed rape varieties grown in Canada, Europe and Chile. This study was undertaken, in part, to resolve organ weight findings from two previous four-week rat feeding studies.

The first study found no differences in liver and kidney weights between the rats fed glyphosate-tolerant canola and those fed control, non-transgenic canola, but this study had to be repeated due to technical problems. In the repeat study, rats fed glyphosate-tolerant canola exhibited a slight (16%) increase in liver/body weight ratio when compared with rats fed control, non-transgenic canola meal. Specifically described in this chapter are compositional data on % fat,% crude protein,% crude fibre,% ash, and sinapines, fatty acid, amino acid and glucosinolate composition in the seed. In addition to the data obtained from harvested seed, representative samples of GT73 and Westar were processed into meal and refined oil and analysed for comparison to accepted values for these components. Analyses of processing fractions and the four-week rat feeding study are considered confirmatory studies since they confirm the primary assessment of agronomic performance and food/feed composition used to establish substantial equivalence of GT73 canola.

## HISTORICAL BACKGROUND OF CANOLA/OILSEED RAPE

Oilseed rape varieties have been cultivated in north-western Europe for centuries, primarily as a source of fuel. While the oil has been used for centuries for cooking in Asia and Mediterranean regions, widespread use of cultivated oilseed rape varieties as a source of human food and animal feed in western countries has occurred more recently. In 1989, rapeseed ranked third in the world production of oilseed crops, surpassed only by soybean and cottonseed. It is grown primarily for the production of oil, which represents more than 40% of the seed weight. The oil is separated from the seed by crushing and processing and is used predominately in cooking oils, margarines and fat. The protein-rich meal remaining behind is employed as a feedstuff for livestock and poultry. The seeds also contain varying levels of the following anti-nutrients: a fatty acid (C22:1) called 'erucic acid' and glucosinolates; a mixture of thioglycosides containing either an aliphatic, aromatic or heteroaromatic side-chain.

Over 100 structural forms of glucosinolates have been identified. Glucosinolates are hydrolyzed to thiocyanates, isothiocyanates, cyclic sulfur compounds and nitriles by myrosinase, an enzyme found in Brassica seed cells. Glucosinolates can also be hydrolyzed or otherwise broken down by microflora in the gastrointestinal tract. Oil from early varieties of oilseed rape was quite high in erucic acids. Oil containing high erucic acid levels (up to 50% of total fatty acids - designated as HEAR oil) has produced myocardial lesions when fed to animals. Oil from more recently developed low erucic acid oilseed rape varieties does not produce myocardial lesions in animals except when fed at high dietary levels to male rats of the Sprague-Dawley strain. Feeding high dietary levels of other vegetable oils to male Sprague-Dawley rats also produces myocardial lesions.

These pathological changes are attributed to the sex- and species-specific sensitivity of this strain of rat to consumption of high levels of vegetable oils. Myocardial lesions have not been observed in other animal species fed oil from low erucic acid oilseed rape varieties; thus the rat is considered an unsuitable model for assessing the safety of rapeseed oil for humans.

The glucosinolates have been traditionally considered to be anti-nutrients since they are metabolized to products such as isothiocyanates which can act as goitrogens. When swine, cattle, poultry and rats were fed varieties of oilseed rape containing high glucosinolate levels, various deleterious effects (e.g. reduced palatability and growth performance, goitrogenicity, liver hypertrophy, etc.) were observed. Commercial processing of oilseed rape reduces the levels of glucosinolates and their metabolites, due in part to inactivation of myrosinase. However, some glucosinolates remain in the meal and metabolites may be formed by microbial flora in the digestive tract.

Nutritionists have encouraged breeders to develop oilseed rape varieties lower in glucosinolates to avoid anti-nutritional effects. Significant progress has been made in recent years to reduce the levels of erucic acid and glucosinolates in oilseed rape through classical breeding approaches.

Canadian varieties of oilseed rape which contain low levels of erucic acid (less than 2% of the total fatty acids present in the oil) and low levels of alkyl glucosinolates (less than 30 μmol/g in defatted meal) may be sold under the trademark designation 'Canola', which is owned by the Canola Council of Canada. Oil derived from the low erucic acid oilseed rape varieties was determined by FDA to be 'generally recognized as safe' (GRAS) for use as a human food.

Processed canola meal has been widely used in Canada as an animal feed for beef and dairy cattle, swine and poultry. The reduced levels of glucosinolates and erucic acid in canola varieties have made it possible for farmers to take advantage of the high protein content of processed canola meal for use in animal feed. The canola varieties are nutritionally superior to older varieties of oilseed rape that had higher glucosinolate and erucic acid content. New varieties of oilseed rape referred to as 'double-zero' have been developed recently in Europe. These are also low in erucic acid and glucosinolate levels compared with canola varieties.

While glucosinolates have been considered to be anti-nutrients, recent evidence suggest that certain glucosinolates (e.g. 3-methylsulfinylpropylglucosinolate) and their metabolites may have beneficial health effects. There is epidemiological evidence that consumption of cruciferous vegetables (e.g. broccoli), which contain a variety of glucosinolates, can reduce the incidence of certain cancers in humans.Considerable research is under way to investigate the effects of glucosinolates on cancer development in animal models and the mechanisms by which glucosinolates act as anti-carcinogens. These studies may help scientists elucidate whether there are human health benefits from consumption of glucosinolates.

## MATERIALS AND METHODS

### Preparation of Canola seed for Compositional Analysis

All plots were planted and managed under normal agronomic practices for canola using seed provided by Monsanto Company and were maintained in strict compliance with all permit requirements of the Canadian Food Inspection Agency. At each location, GT73 of generation R2 or later was planted with a plot of non-modified Westar of the parental variety. The plot sizes ranged from 7.5 $m^2$ to 100 $m^2$. All GT73 samples that underwent compositional analysis were obtained from 100-$m^2$ plots which were completely harvested and subsequently sampled. These plots were not replicated because it was assumed that site-to-site variation in compositional data would be greater than intra-site variation.

Additional data for Westar variety canola were obtained in 1992 (seven locations) and in 1993 (four locations) from the Western Canadian Canola/Rapeseed Recommending Committee Co-Op Field Tests. Certified Westar canola was used as a check (non-genetically modified) variety in these trials. In 1985, Westar was used to plant over 80% of all Brassica napus canola acres in Canada and has been used as a standard in the Western Canadian Cooperative Rapeseed Tests (Co-Op Test) until 1994.

## Compositional Analyses

All analyses were conducted on seed obtained from either the seven field locations in 1992 (Saskatoon, Scott, Melfort, Watrous, Saskatchewan; Minto and Portage la Prairie, Manitoba and Guelph, Ontario) or the four locations in 1993 (Saskatoon, Melfort, Saskatchewan; Minto and LaSalle, Manitoba). Crude protein in defatted meal,% fat (whole seed), glucosinolate and fatty acid composition and sinapine analyses were conducted at Agriculture and Agri-Food Canada's Research Station in Saskatoon (Ag Canada).

In addition, proximate analyses (crude protein, fat, fibre, moisture and ash) of seed from the same field trials were conducted under the US EPA Good Labouratory Practices (GLP) Regulations at Ralston Analytical Labs (RAL) in St. Louis, Missouri. Analyses of meal and refined oil were conducted at RAL and POS (Protein, Oil and Starch Pilot Plant Corporation in Saskatoon, Saskatchewan). The methods used were either validated or based on published methods.

## Processing Study

Processing was conducted on 80-kg samples of GT73 and Westar at POS. Both samples were composites of seed grown at four locations in Canada in 1993. The meal and refined oil samples were produced using procedures developed by POS and equipment suitable for the scale of production. All samples were representative of commercially produced material.

## Statistical Analyses

To determine if any differences in fat, protein, crude fibre, ash, phenylalanine, tryptophan and tyrosine levels between GT73 and

Westar were statistically significant, a randomized complete block design model was used. Each variable in the model Yij = μj + Li + Eij, where Yij is the value obtained for group j at location I, μj is the true mean of group j over all locations, Li is the random effect of location I and Eij is the random effect for group j within location i, was fitted to the data using the GLM procedure in SAS.

SAS ESTIMATE statements in GLM procedure were used to specify the overall means, the mean difference from the Westar mean and the t-test to determine if this difference is statistically significant. For alkyl glucosinolates and erucic acid in GT73, the overall mean, standard error and the 95% confidence interval for the mean were computed using the MEANS procedure in SAS.

The Westar control used in all statistical analyses was planted and grown in an adjacent plot to GT73 at each field location. Compositional data for Westar from the Co-Op Tests conducted in 1992 and 1993 were used for comparative purposes in graphs and tables but were not used in statistical analyses.

## Four-week rat Feeding Study

Processed canola meal was added to rodent diets at a level of 10% (w/w) and fed to rats for four weeks. Ten varieties of canola were tested as follows:

- RU3 (RRC line derived from GT73) and the parental variety (designated as Alliance) grown side-by-side with RU3 in Chile early 1996;
- Canola seed from five commercial varieties grown in different geographical locations across Canada.
- Three commercial oilseed rape varieties grown in Europe in 1995.

The Roundup Ready trait from GT73, the original transformation event, was backcrossed into the Alberta Wheat Pool variety known as Alliance. After three backcrosses, the new variety was selfed several times to create what was identified in the Western Canadian Cooperative Rapeseed Recommending Trials (Co-Op Test) as RU3. There were a total of 15 low erucic acid canola/oilseed rape varieties originally collected for the rat feeding study.

Glucosinolate and compositional data from the lines were reviewed by Dr Sylvie Rabot of INRA Centre de Recherches de Jouy-en-Josas, France. Based on the data, Dr Rabot selected 10 varieties for testing that represent the normal variability in glucosinolate levels that are considered acceptable for commercial use. Seed from the 10 canola/oilseed rape lines was sent to POS for processing. Each lot of seed was divided into duplicate batches of approximately 75 kg of seed so that a total of 20 batches of seed were processed. Random numbers were assigned to each batch so that processing of batches was randomized. The canola seed was flaked, cooked (to inactivate enzymes such as myrosinase), pressed (to remove oil) and the press cake solvent was extracted (hexane) to remove the remaining oil.

The canola meal was further desolventized (toasted) to remove residual hexane. Processing conditions were continually monitored to maintain uniformity across all batches. The toasted canola/oilseed rape meal samples were subsequently analysed at POS using AOAC methods:

- Crude protein;
- Crude fibre;
- Ash;
- Moisture;
- Acid detergent;
- Neutral detergent fibre;
- Nitrogen solubility;
- Myrosinase activity

*Total and individual glucosinolates*: 3-butenyl, 4-pentenyl, 2-hydroxy-3-butenyl, 2-hydroxy-4-pentenyl, 3-methylindol and 4-hydroxy-3-methylindol. The toasted canola/ oilseed rape meal samples were sent to Monsanto's Environmental Health Labouratory (St. Louis, MO) for conduct of the rat feeding study. The toasted canola/oilseed rape meal samples were added to commercial rodent diets (PMI Inc., St. Louis, MO) at a level of 10% w/w). Male and female Sprague-Dawley rats were approximately 5-7 weeks of age at study initiation. Ten (10) male and ten (10) female rats were randomly assigned to each group. There were a total of 21 groups in the study. Twenty (20) groups were fed duplicate batches of either RU3, Alliance or one of the five Canadian or three European varieties of toasted canola/oilseed rape meal. One group was fed only rodent diet (diet control).

Animals were observed twice daily for signs of toxicity. Body weight and food consumption were recorded weekly for each animal. At the end of four weeks, all test animals were sacrificed by carbon dioxide asphyxiation, and the livers and kidneys were removed and weighed. These organs were weighed since it had been reported in the literature that feeding rats canola meal caused increased liver and kidney weights. The in-life phase of the rat feeding study was conducted in conformance with Good Labouratory Practice Standards of the FDA/EPA and Good Labouratory Principles of OECD. The following statistical procedure was used to detect statistically significant differences between the groups. A mixed linear model was used to compare group mean body weights, cumulative weight gain, food consumption, liver and kidneys weights and liver or kidney/body weight ratios. The SAS statistical programme was used to process both the in-life and organ weight data files to extract and compute the data necessary for analysis.

## RESULTS

### Crude Protein and Fat

Measurement of protein and fat levels is a requirement of variety registration for any new canola variety in Canada. Crude protein in the defatted meal and the % fat in the whole seed were determined on seed obtained from Monsanto field trials in 1992 and 1993. Measurement of the % protein and fat in canola seed was conducted at RAL. Other nutrient parameters that were measured included ash, moisture, fibre and carbohydrate (calculated). There were no statistical differences between GT73 and Westar in seed protein or fat content. There were also no differences in the other measured parameters. The proximate values are typical of the published values for B. napus canola.

Protein and fat were also measured at Ag Canada Research Station Labs in Saskatoon using the same methods as those used to evaluate new canola varieties in the Co-Op Field Tests. In 1992, the protein value (% of defatted meal) in GT73 was slightly, but statistically significantly higher compared to Westar; while in 1993, the fat level (% of seed) in GT73 was slightly, but statistically significantly (p0.05) higher compared to Westar. The results are attributed to normal biological variation since increased protein (% of toasted meal) and increased fat (% of seed) were not consistently observed in both 1992 and 1993. Moreover, the increase in fat (% of seed) observed in 1993 (Ag Canada analysis) was not apparent in the aforementioned RAL analysis of fat from canola seed collected from the same 1992 and 1993 Monsanto field trials. A range of values for protein and fat for Westar from the Co-Op Field Tests is also presented. In general, the range of protein and fat values for Westar from the Co-Op Tests (n = 52 in 1992 and n = 87 in 1993) encompass the values for GT73 and Westar from the 1992 and 1993 Monsanto field tests.

### Amino acid Composition

Samples of seed from three of seven Monsanto field trial sites in 1992 and the four Monsanto field trial sites in 1993 were analysed for amino acid composition. Results of these analyses expressed on a per seed basis were very similar for both years. Values obtained from the three sites in 1992 found comparable levels of amino acids for GT73 and Westar. The aromatic amino acids were evaluated more closely because RRC expresses a glyphosate tolerant form of EPSPS which, in addition to endogenous plant EPSPS, is also involved in aromatic amino acid biosynthesis.

Phenylalanine, tryptophan and tyrosine levels in RRC were not different when calculated either on a whole seed or on a per unit protein basis. Values obtained from four sites in 1993 for all amino acids in GT73 and Westar were similar to those found in 1992 (data not shown). Statistical analysis of the aromatic amino acids was again conducted and a small but statistically significant (p0.05) decrease in the level of tryptophan.

Since the differences in tryptophan levels were not consistent between both years, they were attributed to normal biological variation. Moreover, the levels of all amino acids including tryptophan are well within the normal variation expected from canola.

## Fatty acid composition

Seed samples from all seven 1992 Monsanto field trial sites and all four 1993 Monsanto field trial sites were analysed for fatty acid composition at Ag Canada Research Station in Saskatoon. Since the fatty acid levels for GT73 and Westar were similar for both years. Also included in each graph are the ranges of values for the individual fatty acids from the 1993 Co-Op Tests (n = 9 in 1993).

Statistical analysis of the four major fatty acids in canola oil (C16:0, palmitic acid; C18:1, oleic acid; C18:2, linoleic acid and C18:3, linolenic acid) found no differences between GT73 and Westar, with the exception of linoleic acid (C18:2). Linoleic acid in oil from GT73 was statistically significantly (p0.05) lower than Westar in both 1992 and 1993. The magnitude of the difference was small: 0.56% in 1992 (19.23% for GT73 compared to 19.79% for Westar) and 1.01% in 1993. The levels of linoleic acid for GT73, while slightly different from Westar, were still within the range of linoleic acid values from the 1992 and 1993 Co-Op trials for Westar.

The levels for the other fatty acids were also generally within the normal ranges for the Westar variety as determined in the 1993 Co-Op tests. Since erucic acid has been reported to produce myocardial lesions when fed to animals, the Canola Council of Canada (1991) recommended that the level of erucic acid in new canola varieties should not exceed 2% of total fatty acids. The erucic acid levels were well below the 2% limit; and GT73 erucic acid levels were numerically lower than the Westar variety. Statistical analysis of erucic acid levels from the 1992 and 1993 field trials confirmed that at the 95% confidence limit, erucic acid levels for GT73 would not exceed the 2% recommended limit for canola oil.

## Glucosinolates

Glucosinolate levels were analysed in seed from GT73 and Westar in 1992 and 1993 Monsanto field trials. Total alkyl and indolyl glucosinolates (μmol/g defatted meal) are presented along with the range of glucosinolate values from Westar in the Co-Op Test. In both 1992 and 1993, the level of alkyl glucosinolates in GT73, while well below the canola limit of 30 μmol/g defatted meal, was statistically significantly higher than Westar.

There were no statistically significant differences between Westar and GT73 in the total levels of indolyl glucosinolates. Statistical analysis of alkyl glucosinolate levels from the 1992 and 1993 field trials confirmed that at the 95% confidence limit, alkyl glucosinolate levels for GT73 would not exceed the 30 μmol/g defatted meal recommended limit for canola.

A detailed analysis of the levels of individual glucosinolates in GT73 and Westar was carried out to understand what was contributing to the differences in alkyl glucosinolate levels. Most of the increase in alkyl glucosinolates for GT73 was accounted for by hydroxybutenyl glucosinolate, although the levels are well within the expected range for canola (Downey, personal communication) and well below the Canola Council recommended target level of 30 μmol/g defatted meal. This is further supported by analysis of glucosinolates in seed and defatted meal from canola/oilseed rape varieties used in the four-week rat feeding study.

Levels of glucosinolates in seed and processed (defatted) meal from canola/ oilseed rape varieties used in the four-week rat feeding study were also measured at POS. The levels of hydroxybutenyl glucosinolate in RU3 seed were lower than seed from some of the commercial lines of canola/oilseed rape used in the rat feeding study, e.g. 6.8 μmol/g seed for RU3 compared with 4.7 μmol/g seed for Alliance, 3.7-15.7 μmol/g seed for Canadian varieties and 9.6-20.0 μmol/g seed for European varieties.

The most compelling evidence that the aforementioned differences in alkyl glucosinolate levels between GT73 and Westar were not biologically meaningful are based on subsequent glucosinolate analysis of canola lines backcrossed with GT73. Alkyl glucosinolate levels in defatted meal from several backcrossed lines from five different canola seed companies were compared to the parental lines from which they were derived. The levels of alkyl glucosinolates were generally comparable between the parental lines and their progeny resulting from backcrosses with GT73.

### Sinapines

Sinapines are choline esters of naturally occurring plant phenolics derived from cinnamic acid derivatives. Furthermore, these compounds are downstream products of aromatic amino acid synthesis in canola.

They are typically not monitored in new varieties of canola. Their presence in canola meal can impact poultry feed since they are known to render an off-odor in chicken eggs.

### Toasted Meal

Toasted meal samples derived from GT73 and Westar seed were prepared at POS and extensively analysed at RAL. In addition to the proximate and nitrogen solubility analyses, an extensive analysis of minerals and phytic acid was conducted at RAL. The mineral values from GT73-derived meal, were all within the range of literature values and were in agreement with those values obtained from the Westar-derived sample.

In addition, the proximate values for GT73-derived meal were representative of canola and comparable to the values from the Westar-derived sample. These results confirm that meal derived from GT73 is substantially equivalent to meal derived from Westar canola. Refined Oil. Refined oils derived from GT73 and Westar seed were analysed at RAL and compared with standards published in the Food Chemical Codex.

With the exception of fatty acids C22:0, C24:0 and C24:1, the fatty acid levels in GT73 and Westar oil were within the definition of canola oil. Although the levels of these three fatty acids were slightly higher than the Codex standards for both the GT73- and Westar-derived oil samples, they are minor components of the oil and are not considered essential oils for human nutrition.

Thus, no impact on human health would be expected because of slightly higher values than Codex specifications. The slight variation in their levels relative to the Codex Specifications may be a result of breeding lower levels of erucic acid in the Westar variety from which GT73 was derived. The results of the analysis of refined oil derived from GT73 seed confirm that GT73 is substantially equivalent to canola that is currently being marketed. Furthermore, the safety of refined oil from RRC is equivalent to oil refined from non-modified canola.

**Four-week rat Feeding Study**

There were occasional numerical differences between different canola/oilseed rape varieties for some measured parameters, there was generally good agreement between duplicates of the same variety. Processing of canola/oilseed rape to toasted meal produced, on average, a 30% reduction in total glucosinolate levels.

Minimal or no detectable myrosinase activity was detected in the toasted meal samples, indicating that processing conditions were sufficient to inactivate myrosinase and reduce glucosinolate levels. During the course of the four-week feeding study, no test article-related adverse clinical signs were observed. Consistent with the literature, body weights and weight gains were slightly reduced for nearly all of the groups that were fed canola/oilseed rape meal when compared with the diet control.

Food consumption was comparable for males in all groups compared with the diet control. For females, food consumption was slightly lower for some groups fed canola/oilseed rape compared with the diet control. Liver/body and kidney/body weights were slightly increased for most groups fed canola/oilseed rape meal when compared with the diet control pictured.

These increases were due, in part, to the slight reduction in body weights since absolute liver and kidney weights were generally similar to those of diet controls. There were no statistically significant differences (p0.05) in body weight, body weight gain, food consumption, absolute liver and kidney weight, liver and kidney/body weights for male and female rats fed RU3 canola meal when compared with the Alliance variety and the population mean for the commercial canola/oilseed rape varieties that were fed to rats.

The basis of our safety assessment of RRC and the widely accepted basis of assessing the safety of any transgenic food/feed crop is establishing that the initial transgenic line is substantially equivalent to its parental counterpart. The key parameters that should be assessed to confirm substantial equivalence are the agronomic properties of the crop, the compositional analysis of critical nutrients and, if present, important anti-nutrients.

The nutritional or anti-nutritional components that are selected for analysis should be those that would have a meaningful impact on human or animal health should levels be significantly altered in the plant. Regulators in the United States and Europe are collabourating to develop a harmonized list of important nutrients and anti-nutrients/ toxicants for each major food/feed crop that could be analysed to confirm substantial equivalence for a biotech crop. With regard to evaluation of agronomic properties, a number of parameters have been assessed over consecutive growing seasons, including germination; time of flowering; dormancy; over-wintering capacity; pod shattering, pollen fertility time to maturity, seed production; disease and insect susceptibility.

The focus of the food and feed safety assessment of RRC is compositional analysis. Data presented demonstrate that GT73 is substantially equivalent to non-modified B. napus canola. Only two components showed consistent and statistically significant differences between GT73 and Westar: the levels of linolenic acid and alkyl glucosinolates.

In both cases, the differences were of small magnitude and are attributed to the fact that genetic transformation to produce line GT73 represents the selection of a cell from a single seed from the Westar population. Since seed from Westar canola exhibit a range in concentration of nutrients and anti-nutrients, the seed selected for transformation, while within this range, might be expected to exhibit some differences from the Westar population mean for nutrients and anti-nutrients.

The nutrient and anti-nutrient composition data were reviewed in detail by food safety experts in Health Canada and the United Kingdom Ministry of Agriculture, Fisheries and Food who concluded that these differences were not biologically meaningful and approved the oil for human consumption. The alkyl glucosinolate results were also reviewed by plant breeders at the Canadian Food Inspection Agency who developed the Westar variety.

They concluded: 'In our opinion, this is a very minor deviation and one which would be expected when single plants are selected from the heterozygous plant population that constitutes the cultivar Westar'. Subsequent to this testing, the levels of the alkyl glucosinolates in backcrossing events with GT73 and other canola varieties have been monitored by canola seed companies and found to be comparable to the glucosinolate levels in the parental varieties.

To date, a number of varieties of RRC have been approved for cultivation in Canada. In all cases, the fatty acid and glucosinolate composition have been reviewed by experts responsible for canola variety registration in Canada (Western Canada Canola Rape Seed Recommending Committee) and judged to be acceptable for canola. Additional compositional analysis was completed on processing fractions derived from GT73 and Westar seed samples grown at the same locations in 1993. Samples representative of commercially prepared toasted meal and refined oil were obtained and extensively analysed. These confirmatory results demonstrated that processed fractions derived from GT73 and Westar are substantially equivalent. The results of the four-week rat feeding study also confirmed the equivalence of RU3 (RRC line derived from GT73) to other commercial canola/oilseed rape varieties.

The slight reductions in body weights and associated increases in liver and kidney/ body weights observed in all canola/oilseed rape diets have been reported in other rat feeding studies with oilseed rape. There does not always appear to be a dose relationship between the levels of glucosinolates and the changes in relative liver and kidney weights. Smith and Bray found no dose-related increase in liver/body weight as the levels of canola meal in the rat diets were increased incrementally from 10% to 40% (w/w).

Vermorel *et al*. found no dose-related change in liver/body weights for rats fed rapeseed meal with glucosinolate concentrations in meal ranging from 4 to 36 μmol/g. Duplicate samples of toasted canola/ oilseed rape were randomly processed in order to test the variation that might occur during processing. The processing of the batches was carefully controlled to maintain uniformity across all batches as nitrogen solubility indices, an indicator of the degree of processing was comparable across batches.

It was also of interest to assess the responses of animals to feeding duplicate batches of toasted canola/oilseed rape meal. Liver and kidney weights were generally within 1 standard error for most duplicates. However, for 5 out of 40 duplicate measurements, means for liver or kidney/body weights were greater than 1 standard error between duplicate samples.

This points out the random variation that can occur in feeding studies that can sometimes confound the safety evaluation. The absence of any statistical differences in measured parameters between rats fed RU3 (GT73 derived) toasted canola meal at 10% (w/w) in the diet and rats fed processed meal from nine commercial varieties provides confirmatory evidence of the substantial equivalence of RRC to conventional oilseed rape/canola varieties.

This paper outlines a detailed experimental strategy to assess the safety of canola that has been genetically modified to be tolerant to the herbicide glyphosate. The basis for assessing safety is focused on extensive compositional analysis, the results of which lead to the scientific conclusion of substantial equivalence. Stated another way,

RRC is 'as-safe-as' traditional canola. This conclusion is based on the results from over 1000 individual assays conducted over two years in addition to the detailed safety assessment of the introduced proteins and molecular analysis. In addition to establishing the substantial equivalence through field studies (measuring agronomic performance), compositional analyses confirmatory studies, such as the analysis of processing fractions and the four-week rat feeding study, further established that RRC is as acceptable for production of food and feed as non-modified canola.

## Chapter 6

# Mode of Insect Control

The bacterium Bacillus thuringiensis (Bt) is distinguished from other bacilli by the production of large crystals during sporulation composed of one or more insecticidal proteins. These proteins have served as the principal active ingredient of numerous commercial bacterial insecticides used worldwide for more than 40 years to protect crops against caterpillar pests, and more recently to control coleopterous insects and mosquito and blackfly vectors of human and animal diseases.

The first genes encoding insecticidal Bt proteins were cloned early in the 1980s. This led quickly to their use to construct recombinant bacterial insecticides containing novel combinations of these proteins and insect-resistant Bt crops. The first Bt crops became commercially available in the United States in the mid-1990s, and have been widely adopted by farmers despite higher seed costs in comparison with conventional crops.

During the 1999 growing season, farmers in the United States planted over 20 million hectares of insect-resistant Bt transgenic crops, including over 10 million hectares of Bt maize (corn), 1.5 million hectares of Bt cotton, and about 30 000 hectares of Bt potatoes. This acreage is expected to grow to about 15 million hectares of corn and 3 million hectares of cotton within five years, representing, respectively, about a third of the corn and half of the cotton acreage in the United States.

Bt crops are also being grown in China and Argentina, and their potential deployment is being assessed in several other countries. In addition, more than 30 other Bt crops, including rice, sorghum, most major vegetable crops as well as many tree crops used for fruit, nut and fibre production are under development. For farmers, Bt crops offer advantages over conventional crops in that the insecticidal proteins are produced directly by the plants, and continuously during most of the growing season, thereby reducing the material and application costs of using synthetic chemical insecticides.

Growers have reported improved profit margins over conventional varieties averaging $60 to $100 per hectare during the first few years of Bt cotton plantings in the southeastern United States. Consumers and the environment benefit from reductions in the use of synthetic chemical insecticides. For example, it is estimated that the use of Bt cotton in the United States reduced chemical insecticide applications in cotton for lepidopterous pests by 2.7 million pounds in 1999.

This represented a reduction of 15 million insecticide applications (22%) in comparison to the number used in 1995 prior to the use of Bt cotton. These reductions preserve populations of insect natural enemies such as spiders and predatory and parasitic insects, and eliminate exposure of non-target vertebrates, including humans, to these chemicals. The agronomic and environmental advantages of Bt crops are widely recognized in the agricultural community. In fact, the United States Environmental Protection Agency has declared that Bt crops are an "environmental asset".

However, concerns about the long-term impact of their use have been raised by environmentalists and scientists in regard to the development of resistance in insect populations, and the safety of insecticidal proteins produced by Bt crops to humans and other non-target organisms. Important questions have been raised about the safety of Bt crops, but most of the specific test protocols and resultant data relevant to Bt proteins produced by these crops are scarce and are only now being published in refereed scientific journals.

This is because these data were developed primarily to fulfill governmental registration processes required in most countries to sell products, processes which do not require the data to be published in journals. Nevertheless, an extensive database exists in the literature on Bt strains, proteins and products and this information is reviewed here along with data available on Bt crop safety. To address the issue of safety, it is important to understand the role that Bt's insecticidal proteins play in its general biology and the mechanisms by which these proteins exert their toxic effects on insects.

Bt and its insecticidal proteins have a high degree of specificity that accounts for their safety to most organisms, especially in comparison with synthetic chemical insecticides. An understanding of Bt's mode of action, specificity and toxicology, only rarely treated in discussions of safety, is important in assessing any risks. Thus, to place the safety of Bt insecticidal proteins and Bt crops to non-target vertebrates and invertebrates in perspective, I will first summarize our knowledge of the most critical aspects of Bt relevant to understanding this issue, and then deal with the data on the safety of Bt crops.

## THE GENERAL BIOLOGY OF BACILLUS THURINGIENSIS

Bacillus thuringiensis (Bt) is a common Gram-positive, spore-forming bacterium that can be readily isolated on simple media such as nutrient agar from a variety of habitats including soil, water, plants, grain dust, dead insects and insect feces. Its life cycle is simple. When nutrients are sufficient for growth, the spore germinates, producing a vegetative cell that grows and reproduces by binary fission.

The bacterium continues to multiply until one or more nutrients, such as sugars, certain amino acids and oxygen, become insufficient for continued vegetative growth. Under these conditions, the bacterium sporulates to produce a spore and parasporal body, the latter composed primarily of insecticidal proteins. These are commonly referred to in the literature as protein toxins or endotoxins.

Although Bt can be isolated from many environmental sources, and is often referred to as a 'soil bacterium', several features indicate that its principal ecological niche is insects. The original isolations of Bt, for example, were made from diseased caterpillars, and these remain a good source of isolates. More importantly, Bt produces a range of protein toxins and toxin synergists that are very effective at killing certain species of insects, especially larvae of lepidopterous insects, providing a rich substrate for Bt's reproduction.

### Representative Endotoxin-Containing Parasporal Bodies

The principal toxins are protein -endotoxins, the " designating a particular class of toxins, and endotoxin referring to their localization within the bacterial cell after production. Many isolates of Bt also produce the â-exotoxin (a competitive inhibitor of messenger RNA polymerase), insecticidal proteins produced during vegetative growth (Vips), the synergist zwittermicin A, and enzymes such as phospholipases that enhance the activity of the -endotoxins.

In addition, the spore itself can synergize the activity of Bt proteins in some insects, especially those not very sensitive to -endotoxins. In many insect species, especially grain-feeding lepidopterans, Bt reproduces to very high levels after insect death, with millions of spores being produced per insect. Thus, Bt's ecology and reproductive biology suggest that its toxins and toxin synergists evolved to debilitate or kill directly a range of insect species, thereby providing a substrate for reproduction of this bacterium.

### The Diversity of Bacillus Thuringiensis Subspecies

While commonly referred to in the singular as 'Bt', B. thuringiensis is actually a large group of subspecies, all characterized by the production of one or a few parasporal bodies per cell during sporulation. Each parasporal body contains one or more protein endotoxins, typically as crystalline inclusions, and each of these is typically toxic to a limited range of insects. The endotoxins occur in the parasporal body as protoxins, which after ingestion, dissolve and are converted to active toxins through cleavage by proteolytic enzymes in the insect stomach (midgut).

The activated toxins bind to receptors on the midgut microvillar membrane in sensitive insects, lysing the cells and destroying much of the midgut epithelium, causing insect death. In essence, Bt endotoxins are stomach poisons selective for insects and certain other invertebrates. At present, there are more than 70 subspecies of Bt, distinguished from one another by immunological differences in flagellar (H antigen) serotype. Each subspecific name corresponds with a specific H antigen number.

For example, B. thuringiensis subspecies kurstaki is H3a3b, whereas B. thuringiensis subspecies morrisoni is H8a8b. In the literature, the term 'variety' is also used in place of subspecies, as is occasionally the term 'strain'. Because the H antigen serotype-subspecific name often does not correlate with insecticidal properties, acronyms and numbers are often used to designate specific isolates, especially those with important insecticidal properties.

For example, HD1 (isolate number 1 from Howard Dulmage), is used to designate a specific isolate of B. thuringiensis subsp. kurstaki (H3a3b) that produces four major endotoxin proteins and has a broad spectrum of activity against lepidopteran pests. Another isolate of B. thuringiensis subsp. kurstaki (H3a3b) is HD73. This produces only a single endotoxin protein (Cry1Ac) and as a result has a much narrower spectrum of activity against insects than HD1.

Historically, HD1 was the first Bt isolate developed commercially for the control of lepidopterous pests, and remains the most widely used in commercial products today. This isolate has also been the source of much of our knowledge of Bt genetics and molecular biology as well as endotoxin genes used in transgenic bacteria and plants.

Despite the large number of Bt subspecies that have been described, the taxonomic validity of these as well as maintaining B. thuringiensis as a species separate from B. cereus has been in question for many years. At the species level, the primary problem is that the only phenotypic character that clearly differentiates B. cereus from B. thuringiensis is the parasporal body synthesized by the latter species during sporulation. In most Bt subspecies, the information for endotoxin production is encoded on large transmissible plasmids.

When these plasmids are lost naturally or cured from Bt subspecies by growing cells at 42°C, no endotoxins are produced. Cured Bt strains that hosted these plasmids cannot be reliably distinguished from B. cereus. At the subspecies level, phenotypic biochemical differences among many are minor, indicating that many subspecies may not be valid using accepted standards of differentiating bacterial subspecies. In other cases, the differences in biochemical properties and insecticidal activity are so significant that some subspecies could be viewed as distinct species.

From the standpoint of Bt's use as an insecticide, taxonomic studies, especially flagellar serotyping, have aided isolate classification but have proven unreliable as accurate predictors of insecticidal activity. Perhaps the best example of this is found in B. thuringiensis subsp. morrisoni (H8a8b), which includes isolates active against lepidopterous (isolate HD12), coleopterous or dipterous (isolate PG14) insects. Another example of this is found in the occurrence of the 128-kb plasmid that encodes mosquitocidal Cry and Cyt proteins among numerous subspecies of Bt, including israelensis, morrisoni, entomocidus, kenyae and thompsoni.

To summarize the current status of Bt systematics, most molecular evidence is in agreement with more classical biochemical and physiological studies which indicate that B. thuringiensis and B. cereus are the same species, and that the latter becomes the former when it acquires one or more plasmids that express genes for insecticidal proteins. Nevertheless, maintaining Bt as a separate species has practical value because of its insecticidal properties, as does dividing Bt isolates into subspecies based on flagellar antigens.

The latter is useful in cataloguing the more than 20 000 isolates of Bt collected to date. However, to understand the insecticidal properties of an isolate, regardless of its subspecies/serotype or other designation, knowledge of the insecticidal protein genes encoded and expressed is required.

## Insecticidal Proteins of *B. thuringiensis*

Because the â-exotoxin is not permitted in commercial bacterial insecticides in the United States, the two principal active components in commercial Bt preparations are the spore and parasporal body. The endotoxins in the parasporal body account for most of the formulation's insecticidal activity, including initial paralysis followed by death. The lethal effects of these proteins have been known since their discovery in the 1950s. In the early 1980s, shortly after the development of recombinant DNA techniques, it was discovered that Bt -endotoxins were encoded by genes carried on plasmids. This discovery led to a major research effort in many labouratories around the world aimed at understanding the genetic and molecular biology of these toxins. This effort resulted in the cloning and sequencing of numerous Bt genes, over 100 to date, and characterization of the toxicity of the proteins each encodes.

At the time Hofte and Whiteley wrote their review, a wide variety of confusing names and acronyms were being used to refer to Bt's endotoxin genes and proteins. Computer analyses showed that the nucleotide sequences of most of these genes were quite similar. To standardize the terminology, Hofte and Whiteley proposed a simplified nomenclature for naming all insecticidal Bt genes and proteins.

In this nomenclature, the proteins are referred to as Cry (for crystal) and Cyt (for cytolytic) proteins. Though modified recently, this system is still in use today, and is described briefly here along with its modifications and additions. The Cry and Cyt nomenclature developed by Hofte and Whiteley was originally based on the spectrum of activity of the proteins as well as their size and apparent relatedness as deduced from nucleotide and amino acid sequence data, and protein gel analyses.

At that time, with the exception of a 27-kDa cytolytic protein from B. thuringiensis subsp. israelensis, all proteins appeared to be related, and probably derived from the same ancestral protein. Therefore Hofte and Whiteley termed the encoding genes 'cry' genes and the proteins they encode 'Cry' proteins. This designation was followed by a Roman numeral which indicates pathotype (I and II for toxicity to lepidopterans, III for coleopterans and IV for dipterans), followed by an upper-case letter indicating the chronological order in which genes with significant differences in nucleotide sequences were described.

The I and II for lepidopterantoxic proteins also indicate size differences, with the I referring to proteins with a mass of about 130 kDa and the II designating those of 65-70 kDa. Some epithets also include a lower-case letter in parentheses which indicates minor differences in the nucleotide sequence within a gene/protein type. Thus, CryIA referred to a 130 kDa protein toxic to lepidopterous insects for which the first gene (cryIA) was sequenced, whereas CryIVD referred to a 72-kDa protein with mosquitocidal activity for which the encoding gene was the fourth from this pathotype sequenced.

The 27-kDa Cyt1A protein first isolated from B. thuringiensis subsp. israelensis differs from other Bt proteins not only in its smaller size, but also in that it is highly cytolytic to a wide range of cell types in vitro, including those of vertebrates. In addition, it shares no apparent relatedness to Cry proteins. Owing to these differences and its broad cytolytic activity, Hofte and Whiteley referred to this as the CytA protein encoded by the cytA gene. In their revision of Bt gene nomenclature, Hofte and Whiteley listed 38 published gene sequences that encoded 13 different Cry proteins and the single CytA protein. Since their publication, the number of Bt Cry endotoxins has more than tripled, and several new cyt genes have also been described.

As more and more cry genes were sequenced and analysed, it was decided to name genes based on their relatedness as determined primarily from the degree of their deduced amino acid identity. As a result, the nomenclature developed by Hofte and Whiteley was modified in the following manner. The cry and cyt descriptors have been maintained, but the Roman numerals have been replaced with Arabic numbers to indicate major relationships (90% identity), with higher degrees of identity being indicated by following upper-case letters (95% identity), and minor variations of these alleles being designated by lower-case letters, with the parentheses around the latter eliminated.

For example, what was CryIA(c) is now Cry1Ac (and the corresponding gene, cry1Ac), a relatively minor change. However, for some genes/proteins the changes are greater. For example, CryIVD does not cluster with the earlier CryIV (now Cry4) proteins, and is now the taxon Cry11Aa. Although the new designations carry no specific information regarding insecticidal spectrum, primary activity is still inferred because the numbers have been maintained for many of the genes, and there remains a high degree of correlation between relatedness and the spectrum of activity. For example, Cry1 still refers largely to toxicity to lepidopteran insects; Cry2 to lepidopteran toxicity and in some, dipteran activity; Cry3 to coleopteran toxicity, and Cry4 to dipteran toxicity.

## Mode of Action of Cry Proteins

Aside from allergenicity, knowledge of the mode of action and structure of Bt's insecticidal proteins are the two interrelated topics that have most bearing on understanding the safety and risks of insecticides or transgenic crops based on these proteins. Although the spore can play an important role in the pathogenicity of B. thuringiensis to certain insect species, the Cry and Cyt proteins are responsible for the paralysis and death of most target insects.

The three-dimentional structure of three of these is known from X-ray crystallographic studies. Although the mode of action remains to be resolved at a detailed molecular level, these structures along with studies on the effects of the toxins in vivo and in vitro show how Bt proteins kill insects while being safe for most non-target organisms, especially vertebrates. In the typical Bt, B. thuringiensis subsp. kurstaki, for example, the endotoxin crystals dissolve after ingestion upon encountering the alkaline (pH 8-10) juices of the midgut. Dissolution requires the reduction of disulfide bridges that stabilize the Cry molecules in the parasporal crystal. Most Cry toxins are actually protoxins of about 130-140 kDa (e.g. Cry1 and Cry4 proteins) from which an active toxin 'core' in the range of 60-70 kDa is released into the midgut by proteolytic cleavage.

Proper activation results in cleavage of 26-29 amino acids from the N-terminus and about 600 amino acids from the C-terminus. Activated toxin molecules pass through the peritrophic membrane and bind to specific receptors on the apical microvillar brush border membrane of midgut epithelial cells, which lies just beneath the peritrophic membrane.

Binding is an essential step in the intoxication process, and in susceptible insects the toxicity of a particular Bt protein is correlated with the number of specific binding sites (i.e. receptors) on microvilli as well as the affinity of the Bt molecules for these sites. However, binding by itself, even high-affinity binding, does not always lead to toxicity, indicating that insertion and probably post-insertional processing in the midgut membrane is required to obtain toxicity.

For example, in three different studies it has been shown that the Cry1Ac toxin can bind with high affinity to microvillar membrane vesicles from Lymantria dispar, Spodoptera frugiperda and Heliothis virescens, but with little or no subsequent toxicity. This indicates that insertion and likely post-insertional processing of Cry proteins is essential to intoxication. In highly sensitive insect species, the microvilli lose their characteristic structure within minutes of toxin insertion, and the cells become vacuolated and begin to swell.

This swelling continues until the cells lyse and slough from the midgut basement membrane. As more and more cells slough, the alkaline gut juices leak into the hemocoel where, as a result, the hemolymph pH rises by a half unit or more. This causes the paralysis and eventual death of the insect. Though this general picture of the mode of action has been known for some time, the details of the insecticidal process at the molecular level remain unresolved, especially the series of events that occur after the toxin binds to a receptor.

There is good evidence for an immediate influx of potassium and calcium, in response to which the cell takes in water in an effort to balance these cations. To explain this cationic influx, it has been proposed that Bt molecules insert into the microvillar membrane forming transmembrane cation pores. For pores to form, it is postulated that during membrane insertion, the molecules undergo conformational changes that permit insertion.

After insertion the toxin molecules associate to form a pore composed of six toxin molecules. As more and more Cry molecules enter the membrane additional pores form, leading to the influx of cations and water followed by cell hypertrophy and lysis. For a more in depth discussion of the studies dealing with the mode of action of Bt, the interested reader is referred to the reviews by Gill *et al.*, Knowles and Dow and Schnepf *et al.*

## Structure of Cry Proteins

Analysis of cry gene sequences showed that the active portion of the Cry toxin molecule (i.e. essentially amino acids 30-630) contains five blocks of conserved amino acids distributed along the molecule, and a highly variable region within the C-terminal half. The variable region was thought to be responsible for the insect spectrum of activity and experimental evidence from recombinant DNA studies was obtained in support of this hypothesis.

For example, by swapping the highly variable regions of Cry1Aa and Cry1Ac, the insect spectrum of these molecules could be reversed. In addition, binding studies showed that in most cases the degree of sensitivity of an insect to a particular Cry molecule was directly correlated with the number of high affinity binding sites on the midgut microvillar membrane. Studies over the past decade have improved our knowledge of Cry protein molecular biology through determination of the structure of Cry3A, Cry1Aa and Cry2A, identification of regions on the Cry molecule involved in midgut binding and specificity, and identification of several glycoprotein receptors on the insect microvillar membrane.

The structure of the Cry3A molecule was the first to be solved, and similarities in conserved amino acid blocks among Cry toxins and conservation in hydrophobicity indicate that this structure is a good general model for Cry toxins. The solution of the crystal structure of the Cry3A molecule showed that this protein is basically wedge-shaped and consists of three domains.

Domain I is composed of amino acids 1-290 and contains a hydrophobic seven-helix amphipathic bundle, with six helices surrounding a central helix. This domain contains all of the first conserved block and a major portion of the second conserved block of amino acids. Theoretical computer models of the helix bundle of this domain show that after insertion and rearrangement, aggregations of six of these domains could form a pore through the microvillar membrane. Domain II extends through amino acids 291-500 and contains three antiparallel â-sheets around a hydrophobic core. This domain contains most of the hypervariable region and most of conserved blocks 3 and 4.

The crystal structure of the molecule together with recombinant DNA experiments and binding studies indicate that the three extended loop structures in the â-sheets are responsible for initial recognition and binding of the toxin to binding sites on the microvillar membrane. Domain III comprises amino acids 501-644 and consists of two antiparallel â-sheets, within which are found the remainder of conserved block number 3 along with blocks 4 and 5.

The structure resolved by Li *et al.* indicates that this domain provides structural integrity to the molecule. Later site-directed mutagenesis studies of conserved amino acid block 5 in the Cry1Aa molecule suggest that this domain may also play a role in pore formation. Though the structure of the molecule, experiments and computer modeling indicate that binding is attributed primarily to the hypervariable region of this domain, studies by Wu and Aronson also show that positively charged amino acids in domain I are important to binding and toxicity.

Moreover, domain III has also been shown to play a role in receptor binding. Domain I is the pore-forming domain, whereas domain II is responsible for the recognition and binding of the molecule to receptors on the microvilli of the insect midgut epithelium. Domain II therefore accounts for much of the high selectivity and safety of Bt-Cry proteins because mammals and other vertebrates lack receptors for this domain.

Domain III assists in stabilizing the structure of the molecule, and also can play a role in binding. An important aspect of understanding Bt's molecular biology and determining its mode of action at the molecular level is identification of the proteins to which the toxin molecules bind on the microvillar membrane. Sangadala *et al.* contributed to this area with the identification of the first two proteins in the midgut of Manduca sexta to which the Cry1Ac toxin binds.

They identified an amino peptidase of 120 kDa as a major binding protein. More recently, cadherin-like proteins and glycoconjugates have also been identified as receptors for Bt endotoxins. These findings indicate that the initial recognition signals on the midgut of a sensitive insect are not ion channels, but one or more proteins or glycoproteins within or which extend from the microvillar membrane. Thus, if these observations hold for other Cry proteins, the midgut receptors are then essentially docking proteins for the toxins, indicating that the toxins do not directly affect ion permeability by binding to ion channels. These studies of the structure of Bt toxins and their binding properties have identified key regions of the molecule responsible for binding and, putatively, the first binding proteins on the insect midgut epithelium. While further studies are needed to more clearly define the specificity of binding and binding sites, the most enigmatic area of Bt's molecular biology with respect to its mode of action is how the toxin causes toxicity after binding to the membrane.

An understanding of this could provide information that might enable the development of insecticidal Cry proteins for use against insects against which we currently have none, such as cockroaches and grasshoppers.

### Mode of action and Structure of Cyt Proteins

Several Cyt proteins are known, including Cyt1A, Cyt2A, Cyt1B and Cyt2B, though most of our knowledge of these proteins is based on studies of Cyt1A and Cyt2A, and the crystal structure of the latter toxin. These proteins are highly hydrophobic and all have a mass in the range of 21-28 kDa. They share no significant amino acid sequence to identify with Cry proteins, and are thus unrelated. The first Cyt protein, Cyt1A, was identified as a component of the parasporal body of B. thuringiensis subsp. israelensis.

Initially, it was thought that Cyt1A played little role in the toxicity of this subspecies to mosquito and blackfly larvae. However, there is now general agreement that it is not only toxic to these and related flies belonging to the dipteran suborder Nematocera, but that it is important as a synergist of Cry4 and Cry11 toxins, and in avoiding the development of insect resistance.

While rare in comparison to Cry proteins, since the early work on Cyt1A in B. thuringiensis subsp. israelensis, Cyt proteins have been reported from mosquitocidal isolates of B. thuringiensis subsp. kyushuensis (H11a11c), morrisoni (H8a8b), medellin (H30) and jegatheson (H28a28c). As in the case of Cry proteins, Cyt proteins are synthesized during sporulation as protoxins and assembled into crystalline inclusions that make up a portion of the parasporal body. They are always associated with Cry proteins, but do not cocrystallize with these, instead forming a separate inclusion.

The shape of inclusion formed by Cyt proteins varies from hemispherical to angular because they assemble in a spherical parasporal body simultaneously along with inclusions formed by the other Cry proteins with which they occur. However, when produced alone in a recombinant B. thuringiensis, the Cyt1A protein forms a bipyramidal crystal with 12 faces. Based on studies of Cyt1A, after ingestion the 27.3-kDa protoxin molecules dissolve from the crystal under the alkaline conditions present in the midgut of susceptible insects such as mosquito larvae.

Proteolytic enzymes then activate the protoxin by cleaving amino acids at both the C- and N-terminus, releasing an active toxin of 24 kDa. This molecule passes through the peritrophic membrane where it then inserts into the microvillar brush border membrane of midgut epithelial cells. Unlike Cry molecules, it appears that Cyt proteins do not require a glycoprotein receptor for binding. Instead, Cyt proteins have a high affinity for the lipid portion of the membrane, specifically for unsaturated phospholipids such as cholesterol, phospholipidcholine and sphigomyelin.

After binding to the microvillar membrane, the cells hypertrophy and subsequently lyse. The specific mode of action of Cyt toxins at the molecular level is not known. The leading hypothesis is that Cyt molecules insert into the microvillar membrane and then assemble into clusters of as many as 18 molecules to form pores that act much like those proposed for Cry toxins. Lysis then results from the influx of cations and water. However, evidence has also been provided that Cyt molecules do not form pores, but rather act like detergents, binding to and perturbing the lipid bilayer and disrupting the structural arrangement of membrane proteins, which would also cause lysis.

Unlike Cry proteins, Cyt proteins exhibit cytolytic activity in vitro against a wide range of invertebrate and vertebrate cell types, though at higher concentrations. However, toxicity is not observed when ingested in vivo by most insects tested, including larvae of lepidopterous species, non-nematoceran flies such as houseflies and fruitflies, and vertebrates such as mice.

The mechanism of this in vivo specificity is not known, but may involve specific combinations of unsaturated phospholipids. Though Cyt proteins have been thought to be only toxic to nematocerous dipterans, it has recently been shown that at least one beetle species, Chrysomela scripta, is sensitive to the Cyt1A protein. An important property of Cyt proteins, based primarily on studies of Cyt1A, is their ability to synergize the toxicity of the Cry proteins with which they occur.

For example, combination of Cyt1A with the Cry4A, Cry4B or Cry11A proteins with which it occurs in the parasporal body of B. thuringiensis subsp. israelensis results in toxicities three- to fivefold higher than the specific toxicity of Cyt1A or any of the Cry proteins alone. The mechanism of this synergism is not known but probably involves cooperativity of the Cyt and Cry proteins in binding to and/or inserting into the microvillar membrane.

## Bacterial Insecticides and Transgenic Crops

More than 30 Bt formulations are on the market in the United States, Europe, Asia and South America. Most of these are complex mixtures of sporulated cells produced by large-scale fermentation and formulation ingredients (stabilizing agents, spreaders, stickers, fillers).

The fermentation products contain spores and an array of toxins found in either B. thuringiensis subsp. kurstaki, which is used to control a wide range of lepidopteran pests, or B. thuringiensis subsp. israelensis, used to control the larvae of numerous mosquito and blackfly species. Both owe their success to their broad spectrum of activity and the absence of any widespread insect resistance after many years of use.

These characteristics are due to the complex composition of the parasporal body, which in each case contains four major proteins, and to the moderate frequency of usage. With respect to Bt transgenic crops, those on the market are based on full-length or truncated Cry proteins. These crops were constructed by transforming the plants with Bt cry genes modified to optimize their expression in plants by changing the codons to reflect plant codon usage.

All Bt crops currently on the market produce only a single Cry protein, Cry1Ab, Cry1Ac or Cry3A. However, crops which produce two Cry proteins, such as Bt cotton producing Cry1Ac and Cry2Ab, are being tested in the field. In addition, a variety of insect-resistant crops which produce Bt non-Cry or non-Bt insecticidal proteins are under development, and submissions for regulatory approval have been made in the United States and Australia.

### Safety of *Bacillus thuringiensis* - General Considerations

Knowledge of Bt's biology, especially the information on the mode of action and structure of its -endotoxin proteins, forms a sound foundation for understanding and evaluating Bt safety. From these studies, it can be seen that these proteins have a series of 'built-in' levels of specificity that should provide Bt bacterial insecticides and Bt crops with a high degree of safety for vertebrates and most non-target invertebrates. For Bt insecticides based on Cry proteins, these specificities include the following:

- Endotoxin crystals and spores must be ingested to have an effect; there is no 'contact' activity, as occurs with chemical insecticides. This is the reason sucking insects and other invertebrates such as spiders and mites are not sensitive to Bt.

- After ingestion, the endotoxin crystals of Bt subspecies used to control lepidopterous pests and mosquitoes and blackflies require alkaline conditions, typically a pH in the range of 8-11 in the midgut lumen, to be solubilized in a form conducive to activation by midgut proteases. Under the highly acidic conditions that exist in the stomachs of many vertebrates, including humans, the endotoxin crystals may dissolve, but most of the solubilized protein is rapidly degraded, often within minutes, by gastric juices to non-toxic peptides.
- Once in solution in the insect midgut lumen, the toxin must be properly activated through cleavage by midgut proteases. Improper cleavage results in inactive toxin.
- Once activated, the toxin must bind to 'receptors' on midgut microvilli. Most chewing insects that ingest toxin crystals, even if they have alkaline midguts, including many lepidopterans, do not have the appropriate receptors, and thus are not sensitive to the toxins. Even insects sensitive to one class of Bt proteins, such as lepidopterans sensitive to Cry1 proteins, are not sensitive to the Cry3A proteins active against coleopterans, one reason being they lack the receptors for these.
- After binding to a midgut receptor, the toxin must enter the membrane, change conformation in the process, and oligomerize to form pores that will be toxic.

With respect to level E, at present, the specific conformational changes that must take place to exert toxicity are not known. It is known, however, that high-affinity irreversible binding can occur in some insects, yet not lead to toxicity. This implies that a specific type of processing (i.e. another level of specificity) is required for toxicity that occurs as or after the toxin inserts in the membrane.

In Bt crops, only a portion of the second level (i.e. B) of the first five levels of specificity has been circumvented. When synthesized in plants, full-length and truncated Cry proteins do not form crystals, and even if quasi-crystalline inclusions do form, toxin remains in solution within the plant cells.

Nevertheless, whether full-length or truncated protoxins, Cry proteins produced by Bt crops must still be properly activated after ingestion, and must successfully meet the criteria for binding and membrane insertion by levels D and E to be toxic. Furthermore, with the one exception of the Cry9C protein, which was engineered to resist rapid proteolytic cleavage, Bt proteins produced in Bt crops are rapidly degraded under conditions that mimic the mammalian digestive system. Therefore, most of the inherent levels of specificity that account for the safety of Cry proteins used in commercial bacterial insecticides apply to these same proteins when used to make Bt crops resistant to insects.

For Cyt proteins, which are not used in any Bt crops at present, the specificity in vivo has the same levels or stages in A-C. However, membrane insertion does not appear to require a protein receptor, but rather the presence of specific unsaturated fatty acids in the lipid portion of the microvillar membrane. Lastly, an important concept of evaluating safety is to consider the route by which an organism is likely to encounter a toxin.

Even though pulmonary (inhalation) and intraperitoneal injection studies are done with microbial Bt insecticides and proteins, their normal route of entry by target and non-target organisms is by ingestion. This is even more so for Bt proteins produced in Bt crops.

## Safety Testing for Bacterial Insecticides

In addition to their insecticidal efficacy, a major impetus for using Cry proteins in Bt crops was their long history of safety to non-target organisms, especially to vertebrates. The most important levels of Bt endotoxin specificity, i.e. activation, binding and membrane insertion, apply equally to evaluating the safety of Cry proteins whether used in Bt crops or bacterial insecticides. Therefore, the tests and data that support a very high degree of safety for bacterial insecticides containing Cry proteins are relevant to assessing the safety of Bt crops.

Extensive testing has been and remains required to meet rigorous safety requirements established by governmental agencies such as the US Environmental Protection Agency. Use of data from these tests is valid as a major approach to evaluating Bt crop safety, especially considering that many hundreds of safety tests have been conducted over several decades to register numerous bacterial insecticides based on different subspecies of Bt.

The principal Bt subspecies evaluated in these tests have been B. thuringiensis subsp. kurstaki, B. thuringiensis subsp. israelensis, B. thuringiensis subsp. aizawai and B. thuringiensis subsp. morrisoni (strain tenebrionis). The materials evaluated have been the active ingredients, i.e. sporulated cultures containing spores and crystals of Cry and Cyt proteins, as well as formulated products. Among the materials tested are all of the Cry proteins used in commercial Bt crops currently on the market. In determining what types of tests should be done to evaluate the safety of bacterial insecticides, early tests were based primarily on those used to evaluate chemical insecticides. However, the tests have evolved over the decades and are now designed to evaluate the risks of Bt, specifically the infectivity of the bacteria and toxicological properties of proteins used as active ingredients.

The tests are grouped into three tiers, I-III. Tier I consists of a series of tests aimed primarily at determining whether an isolate of a Bt subspecies, as the unformulated material, poses a risk if used at high levels, typically at least 100 times the amount recommended for field use, to different classes of non-target organisms. The principal tests include acute oral, acute pulmonary (inhalation) and acute intraperitoneal evaluations of the material against different vertebrate species, with durations from a week to more than a month, the length depending on the organism. In the most critical tests, the mammals are fed, injected with, and forced to inhale millions of Bt cells in a vegetative or sporulated form.

Against invertebrates, the tests are primarily feeding and contact studies. Representative non-target vertebrates and invertebrates include mice, rats, rabbits, guinea-pigs, various bird species, fish, predatory and parasitic insects, beneficial insects such as the honeybee, aquatic and marine invertebrates, and plants. If infectivity or toxicity clearly results in any of these tests, then the candidate bacterium would be rejected. If uncertainty exists, then tier II tests must be conducted. These tests are similar to those of tier I, but require multiple consecutive exposures, especially to organisms where there was evidence of toxicity or infectivity in the tier I tests, as well as tests to determine if and when the bacterium was cleared from non-target tissues.

If infectivity, toxicity, mutagenicity or teratogenicity is detected, then tier III tests must be undertaken. These consist of tests such as two-year feeding studies and additional testing of teratogenicity and mutagenicity. The tests can be tailored to further evaluate the hazard based on the organisms in which hazards were detected in the tier I and II tests. To date, none of the registered bacterial insecticides based on Bt have had to undergo tier II testing. In other words, no moderate or significant hazards or risks have been detected with any Bt subspecies against any of the non-target organisms studied so far.

As a result, all Bt insecticides are exempted from a tolerance requirement, i.e. a specific level of insecticide residue allowed on a crop just prior to harvest. Moreover, no washing or other requirements to reduce levels consumed by humans are required. In fact, Bt insecticides can be applied to crops such as lettuce, cabbage and tomatoes just prior to harvest. It is important to realise that such a statement cannot be made for just about any synthetic chemical insecticide. This does not mean that registered bacterial insecticides do not have any negative impacts on non-target populations or do not harm non-target organisms, but rather that these materials do not pose a significant risk to populations of these organisms or the environment.

## SAFETY OF NON-TARGET INVERTEBRATES

The concept of a non-target organism is a relative one and therefore requires some clarification. The term non-target organism generally refers to organisms outside the main target group. For example, with most organophosphate, carbamate and pyrethroid insecticides, because they often are capable of killing many different types of insects as well as other invertebrates such as spiders and crustacea, non-target organism usually refers to non-insect species.

With Bt insecticides, owing to their high selectivity, the definition of a non-target organism typically is much broader and includes all insects outside the taxonomic order to which the primary target insects belong. Bt insecticides are so specific, even against insects, that their spectrum of activity is typically identified in a very narrow manner, such as 'lepidopteran-active', 'dipteran-active' or 'coleopteran-active'. Even then, Bt insecticides are so specific that a Bt subspecies generally characterized as 'lepidopteran-active' may be highly toxic to some lepidopteran species, but have only low or no toxicity to others.

This point can be illustrated with the HD1 isolate of B. thuringiensis subsp. kurstaki (Btk), the isolate used widely in commercial formulations to control lepidopteran pests. Btk is highly toxic and very effective against larvae of the cabbage looper, Trichoplusia ni, a common pest of vegetable crops, but typically exhibits poor activity against the beet armyworm, Spodoptera exigua, another important caterpillar pest that belongs to the same taxonomic family. This is because none of the toxins produced by Btk (Cry1Aa, Cry1Ab, Cry1Ac and Cry2A) are very toxic to Spodoptera species.

For this reason, the product XenTari (Valent BioSciences) based on B. thuringiensis susbp. aizawai, which produces a protein (Cry1Ca) of moderate toxicity to Spodoptera species, was developed for control of Spodoptera species. With respect to the 'dipteran-active' B. thuringiensis subsp. israelensis, this subspecies is only highly toxic to species of the suborder Nematocera ('long-horned flies' meaning flies with long antennae), a subdivision of the fly group, which, in addition to mosquitoes and blackflies, contains the mushroom flies, craneflies and chironomid midges. Flies such as houseflies, horseflies, stable flies, and the many types of fruitflies are not sensitive to toxins or spores of this subspecies. A high degree of specificity and hence safety is attributed to each Bt insecticide, meaning that a Bt subspecies that serves as the active ingredient is limited to being toxic primarily to the insect species of only one taxonomic order.

Nevertheless, this would still mean that many non-target species of this order would be sensitive to the Bt endotoxins by the normal route of entry (i.e. ingestion). Thus what we consider a pest is an arbitrary concept as opposed to one based on taxonomy. This has led to considerable misunderstanding about the effects of 'lepidopteran-active' Bt subspecies used as insecticides, or the proteins derived from these that are used in Bt crops. An isolate like HD1 of B. thuringiensis subsp. kurstaki has a broad host range against lepidopteran species, due primarily to the four insecticidal proteins it produces.

Therefore, when used in the field it will be capable of killing larvae of target as well as certain non-target lepidopterans. Among the targets are larvae of many moth species, especially those of the family Noctuidae (e.g. the corn earworm, the cotton budworm and bollworm, and the cabbage looper). Among the non-targets in certain geographical areas are the larvae of non-pest lepidopterans including larvae of the monarch butterfly, and many other species of moths and butterflies, some of which are endangered species.

This can pose a dilemma for farmers as well as the governmental agencies, both regulatory agencies and local governments, in making decisions about the effects of Bt insecticides, and now Bt crops, on non-target organisms. The choices in the case of endangered species are particularly difficult, and can be illustrated with control of the gypsy moth, Lymantria dispar.

Most communities in the eastern and mid-western regions of the United States have banned the use of synthetic chemical insecticides to control this pest in forests and residential areas. Bt is used despite the protests of many environmentalists. The reason is that Bt represents the most environmentally compatible choice, even for endangered lepidopteran species.

If synthetic chemical insecticides were used, their impact on all non-target invertebrates would be much more devastating than the use of Bt. If Bt is used, populations of the gypsy moth are controlled, at least temporarily, but there are also losses of sensitive non-target lepidopterans in the treated habitats.

However, the option of not treating with Bt is worse in that gypsy moth larvae periodically defoliate most plants, thereby eliminating the food sources of all leaf-eating insects. Alternatively, at least some of the larvae of the non-target species will survive the Bt treatment, whereas few, if any would survive the use of a chemical insecticide. With respect to specific evaluations of Bt insecticides against non-target invertebrates, there have been numerous studies in the labouratory as well as in field situations under operational pest and vector control conditions.

Literally thousands of tons of Bt insecticides have been applied in the environment over the past four decades, and the overall record, especially considering the amounts applied, is one of remarkable safety. For more specific details and references to the extensive primary literature on non-target studies, the interested reader is referred to the recent comprehensive text by Glare and O'Callaghan, as well as review articles by Couch and Foss, Dejoux and Elouard, Lacey and Mulla, Meadows, Melin and Cozzi, Mulla and Vinson.

Bacterial insecticides based on different subspecies of Bt have been tested extensively in the labouratory against non-target invertebrates to meet registration requirements, and have also been evaluated in field situations to assess effects of formulated products under operational conditions. Both short-term (i.e. from a few days to several weeks) as well as long-term studies of more than a year have been conducted. In the labouratory studies, doses used to evaluate the effects on non-targets are typically as much as 1000-fold the amount that these invertebrates would encounter in the field, and in many cases the doses are much higher.

Representative non-target invertebrates that have been studied include earthworms and microcrustacea such as daphnids and copepods that make up much of the zooplankton in treated areas. In addition, insects tested have included non-target Coleoptera (beetles), Diptera (flies), Neuroptera (lacewings), Odonata (dragonflies and damselflies), Trichoptera (caddisflies) and Hymenoptera (parasitic wasps), especially species that constitute the major predator and parasite groups that attack the insect pests or disease vectors that are the targets of the Bt applications.

Larvae and adults of beneficial insects such as the honeybee, Apis mellifera, are also tested. In testing Bt products used against caterpillar pests, more emphasis has been placed on evaluating the effects on terrestrial non-target invertebrates. However, because these products can drift or be washed into streams and ponds, many aquatic invertebrates have been tested in labouratory studies and in natural habitats. In the case of B. thuringiensis subsp. israelensis, used to control mosquito and blackfly larvae, greater emphasis has been placed on evaluating the effects on aquatic non-target insects and other arthropods. Summaries of these results and those of other studies carried out over the past 30 years show virtually no adverse direct or indirect effects, especially long-term effects, of Bt or formulated products of Bt on non-target populations. The obvious exceptions are non-target species that are closely related to the target pests or vectors, or insects such as endoparasitic hymenopteran species that require the target lepidopteran pests as hosts. But even these are not affected in some cases.

Even in 'forced' feeding studies, the Bt subspecies did not have an effect on insects or non-target invertebrates, such as shrimp, that were outside the order of insects designated as the target group. In some of the earliest studies, effects were seen on earthworms and flies. But these early studies were conducted with strains that may have had the â-exotoxin, which has not been permitted in commercial formulations for decades.

In cases where the effects of Bt on non-target populations have been monitored under field conditions, the effects on non-target organisms were much less than those resulting from the use of chemical insecticides. The effects of Bt therefore must be viewed from the perspective of the consequences of using alternative control technologies. An appropriate example is the use of B. thuringiensis subsp. israelensis (Bti) in the Volta River Basin to control the larvae of Simulium damnosum, the blackfly vector of onchocerciasis, a blinding eye disease of humans.

The onchocerciasis control programme is sponsored by the World Health Organization and United Nations Development Programme. After more than a decade of intensive use, it was concluded that Bti was of 'only the slightest of hazards' to any of the non-target organisms tested. More specifically, when Bti formulations were applied to rivers, the 'drift' of invertebrates (i.e. the target and non-target invertebrates found floating in the rivers and presumably killed or disturbed by the application) increased two- to three-fold in comparison with untreated rivers.

However, when chemical insecticides were applied under similar ecological conditions, the drift increased 20- to 40-fold. In other words, the application of chemical insecticides was approximately ten times more detrimental to the non-target invertebrate populations than the use of Bti. In addition to the much greater impact of the chemical insecticides on non-target invertebrates in the rivers, the blackfly population began to develop resistance to these chemicals. Replacement of the latter with Bti-based insecticides, to which no resistance has developed, during the drier periods of the year ensured the success of this programme, and allowed large fertile areas of the river valleys in West Africa to be returned to productive agriculture.

## SAFETY OF BT TO MAMMALS

Commercial formulations of Bt insecticides are used to control caterpillar pests on many vegetable crops and in forests near residential areas. They are also used to control mosquito and blackfly larvae in rivers, reservoirs and other bodies of water used for human consumption. As a result, and because other bacilli such as B. cereus and B. anthracis can infect humans, the Bt subspecies that serve as the active ingredients as well as the formulations made from these have been tested extensively for their safety to vertebrates, especially mammals.

These tests are designed primarily to assess whether the bacteria have the potential to infect mammals, and whether the endotoxins are toxic when administered through different routes of entry. In addition, eye and dermal irritability tests have also been conducted. In some cases, preparations have even been tested on human volunteers. Standard tests required by governmental agencies include feeding studies, pulmonary studies, and intraocular and intraperitoneal inoculation studies.

Standard test animals include mice, rats and guinea-pigs, but non-mammals including chickens and other birds as well as fish have also undergone similar tests. A typical test involves a replicated series of trials in which mice and/or rats are injected intravenously or intraperitoneally with the candidate preparation (one million Bt cells per mouse, 10 million cells per rat).

The test animals are then observed for several weeks for infection or toxic reactions. Attempts are made during and at the end of the study to isolate bacteria as a measure of the rate at which inoculated cells grow in, or are cleared from, the test animals. Successful isolation, however, is not an indication of infection because this can be due to the persistence of spores within tissues, as noted by Siegel and Shadduck.

Therefore, criteria for infection are isolation of vegetative cells and an increase in the Bt population. Tests have also been done where large numbers of cells have been injected into the mouse brain. The results of numerous feeding, ocular, pulmonary and interperitoneal injection tests conducted with B. thuringiensis subsp. kurstaki, B. thuringiensis subsp. israelensis and B. thuringiensis subsp. morrisoni (strain tenebrionis) against mice and rats have shown no evidence of infection or toxicity to the test animals. Even in immunodeficient mice, no infection was detected after interperitoneal injection of as many as 10 million colony forming units. In pulmonary tests, B. thuringiensis subsp. israelensis cells were cleared from the lungs within four days. However, in the intraperitoneal tests, viable Bt cells could be isolated from mice up to 10 weeks after inoculation. Nevertheless, there was no evidence of infection,or toxicity after 10 weeks. There are limits to such aggressive testing procedures where large doses of microorganisms are administered.

For example, Siegel and Shadduck found that intracerebral injection of B. thuringiensis subsp. kurstaki and B. thuringiensis subsp. israelensis caused high mortality when 10 million cells were injected into the brains of weanling rats. But even in these tests, most rats injected with one million or fewer cells survived and showed no signs of infection. Intraperitoneal injection of mice with 100 million colony forming units did cause high mortality with some isolates of Btk and Bti, but the results were not consistent, and lower amounts of 10 million cells caused no illness or mortality.

Similar tests with B. subtilis, which does not produce Cry toxins, produced comparable results. In contrast to these results, there is one study in which an unidentified Bt Cry protein putatively caused abnormalities in the ileum of mice fed potato treated with Bt. The micrographs presented in this paper showed clear evidence of cell damage. But the cause of this damage remains unclear. The strain was not characterized with respect to endotoxin content, and it is possible that the damage was due to $\alpha$-exotoxin, which is present in many Bt isolates and known to damage vertebrate intestinal mucosa.

This study is also suspect because the authors erroneously attribute cytolytic proteins and properties to B. thuringiensis subsp. kurstaki that are not characteristic of this subspecies, but rather are properties of B. thuringiensis subsp. israelensis. In addition, in other studies in which the ileum of mice was examined using immunocytochemical methods, no receptors for the Cry1Ab protein were found.

Only a few tests have been conducted on human volunteers, and these were feeding studies in which each individual was fed 1 g of Bt (10 billion spores along with endotoxin crystals) per day for three consecutive days. This is at least 1000 times the dose a human would consume by eating fresh vegetables immediately after treatment with a Bt formulation at the recommended rate. Yet no infections or illness occurred. In addition, there have been no confirmed human infections or deaths during the 40 years that commercial preparations have been used for pest and vector control around the world. This statement cannot be made for chemical insecticides, where hundreds of cases of illness and 5-10 deaths occur yearly as a result of the misuse of methyl parathion alone. An overwhelming amount of evidence indicates that Bt products, owing to their highly specific mode of action, cannot be 'misused' in the sense that they could accidentally cause illness or death.

There have been a few reports of humans 'infected' with Bt, or Bt causing infections in domesticated animals. However, aside from being extremely rare, these reports have not been substantiated, and the presence of Bt may have been due to persistence rather than infection. McClintock *et al.* concluded that Bt proteins were not involved in any of these rare reports. Considering the widespread use of Bt, and its simple nutritional requirements for growth, if it had even marginal potential as a pathogen for humans and domesticated animals it should have been isolated as the cause of infection much more frequently.

It is appropriate to note that the consumption of fresh vegetables, especially organically grown lettuce, cabbage, broccoli and tomatoes, even if washed, results in routine human consumption of viable Bt spores and endotoxin crystals. In addition to being allowable as a spray on crops just prior to harvesting, Bt is common in the environment.

## SAFETY TESTING

The genetic engineering of crops to make them resistant to insects through the production of Bt Cry proteins has raised concerns about the safety of these crops to non-target invertebrates and vertebrates, especially their safety to humans. Proponents of Bt crops argue that the safety of Bt crops has been established by direct feeding studies as well as by the 40-year safety record established by Bt insecticides. Moreover, they argue that Bt crops are even better because they typically contain only one Cry toxin and thus lack the other components of Bt insecticides such as spores, fermentation residues and formulation additives.

Alternatively, opponents of Bt crops counter that their safety concerns are valid because Bt crops produce Cry proteins in a form that is modified from what the bacteria produce. For example, Cry proteins do not form crystals in plants, but remain in solution, thereby eliminating the dissolution step from the different levels of specificity. In addition, they argue that Bt proteins are synthesized in a truncated form making them already active, and that it is possible that Cry proteins undergo processing in plants, such as glycosylation (which has not been demonstrated), that could make them toxic or allergenic to mammals. The opponents of Bt crops state therefore that the safety record of Bt insecticides cannot be relied upon to validate the safety of Bt crops. The concerns raised by the opponents of Bt crops have been widely reported in the popular press, creating a public backlash against Bt crops, especially in several European countries. The backlash has been minimal in the United States and countries such as Canada, China and Argentina. However, the finding that sufficient quantities of Bt maize pollen can kill larvae of the popular monarch butterfly under certain labouratory conditions was reported in most newspapers 1999.

This resulted in a higher level of concern by the public about Bt crops, at least about their potential effects on non-target organisms. These concerns have resulted in additional scrutiny of existing tests and data used by scientists and governmental agencies to determine the safety of Bt crops, as well as additional investigations under typical field conditions. Based on a recent series of studies conducted under field conditions, it was concluded that pollen from Bt maize would have only a negligible effect on monarch populations.

Before reviewing the types of studies and scientific evidence used to access Bt crop safety, some additional background is provided here about the forms of Cry proteins produced by crops. Even if in solution, the toxins still must undergo activation and meet the other criteria for toxicity summarized earlier. Most crops currently on the market produce full-length Cry or partially truncated proteins, typically Cry1Ab or Cry1Ac. In crops that produce truncated Cry proteins, the truncated version lacks a major portion of the C-terminal half of the molecule.

It has been assumed in some studies that this truncation results in the production of active toxin in plants. Activation of the toxin by midgut proteases also cleaves 26-29 amino acids from the N-terminus of the molecule, so even the truncated Cry proteins produced by plants are not fully activated, unless this fragment has been removed by plant proteases. It should also be realised that even in bacterial insecticides, some activated toxin is present due to the action of bacterial proteases produced during sporulation.

Most of the toxin in crystalline form is protected from activation, but toxin in solution at the time of lysis is subject to activation. Thus the safety tests carried out with the various Bt subspecies and formulated products based on these would have included activated Cry proteins. If these were toxic to non-target organisms, this toxicity should have been detected in the safety tests conducted over the past 3-4 decades, especially considering the high doses used in these tests.

The possibility remains that some plants may modify Cry proteins in a way that the bacteria from which they were derived do not. The developers of Bt crops have examined Bt proteins produced by various Bt crops for post-translational modification, including glycosylation, phosphorylation and acetylation. No indication has been found that these proteins have been modified post-translationally.

Even if post-translational modification of Bt proteins is eventually found to occur in some Bt crops, there is no reason to assume that these modifications would make a Bt crop less safe to mammals or non-target invertebrates. Many plant proteins, for example, seed storage proteins that are widely consumed by humans, are glycosylated, yet cause no harmful effects. Furthermore, if post-translational modifications were to occur, owing to the complex mode of action of Cry proteins, they are just as likely to interfere with the toxicity of Cry proteins to target insects, possibly making the crops ineffective.

Present Bt crops represent an early phase of a new technology, and it is easy to exaggerate their potential benefits and shortcomings. Given the 40-year safety record of Bt insecticides along with the well-accepted empirical methods for testing the safety of chemical insecticides, bacterial insecticides and drugs, it is appropriate that a combination of prior studies and empirical methods be used to establish the safety or lack thereof of Bt crops. Over the past few years, studies of Bt crop safety using empirical methods have begun to appear in the scientific literature.

These studies have examined the effects on non-target invertebrates and vertebrates including mammals in the labouratory and field. Under operational growing conditions, without exception, these studies show that Bt crops have no significant adverse consequences for non-target invertebrate populations, and if anything their use is beneficial because the amount of broad-spectrum chemical insecticides used is reduced.

Replacement of chemical pesticides with Bt crops provides better protection of beneficial insect populations due to the much greater specificity of Bt proteins. The labouratory studies also provide a variety of evidence that Bt crops are safe for human consumption.

## SAFETY OF CROPS TO NON-TARGET INVERTEBRATES

Cry proteins produced by transgenic plants are not easily extractable in the amounts that would be required for studies designed to test the effects on non-target organisms. Thus, the effects are currently assessed by feeding test species Cry proteins produced in either Escherichia coli, a Bacillus species, or on various Bt crop tissues such as leaves or pollen. Tests of Cry proteins produced in E. coli are similar to those used to evaluate these proteins when produced by B. thuringiensis, except that in many cases an activated form of the toxin is used to produce what could be considered a 'worst case' risk assessment. To complement labouratory studies, several field studies have been conducted in which non-target insect populations were monitored on Bt crops, mainly Bt maize and Bt cotton, throughout the growing season. Most of the labouratory studies have been performed in the United States, where a complex of non-target organisms serves as a standard group for which results are accepted by the US Environmental Protection Agency.

The standard test invertebrates have included a range of terrestrial and freshwater aquatic organisms generally considered beneficial. These typically are larval and/or adults of one or more of the following organisms: the honeybee, parasitic wasps, predatory ladybird beetles and lacewings, 'soil-dwelling' springtails, earthworms, and as a representative of a fresh-water aquatic crustacean, a daphnid. The non-mammalian non-target vertebrates have typically been a bird and a fish, respectively, the bobwhite quail and the channel catfish.

In these tests, the non-target organisms were typically exposed to or fed amounts of toxin that were in the range of at least a hundred to several thousand times the amount they would be exposed to or consume under natural conditions. In such tests, when no effects are observed at the highest dose or rate tested, this amount is referred to as the no-observed-effect-level (NOEL). For a crop like Bt maize, the amount of Cry protein in a maturing field is estimated to be about 500 g per hectare, and thus the test levels are adjusted to ensure a dose at least 1000 times this level.

Most of the studies conducted so far have been short-term studies, lasting from several days to a few weeks. In addition to these studies aimed at testing high levels of exposure to Cry proteins under situations that resemble a natural situation, there have been tests where non-target invertebrates have been forced to feed on Bt crops. In Italy populations of the aphid Rhodopalosiphum padi were force-reared on Bt maize (Cry1Ab).

Then its natural predator, the green lacewing, Chrysoperla carnea, was fed on these aphids. In comparison to control populations, no adverse effects were noted on either the aphid or lacewing populations with respect to survival and fecundity. In contrast to these results, other labouratory studies have shown that force-feeding can result in mortality to certain non-target insect species.

For example, it was shown that immature C. carnea fed on prey that had been fed Bt maize (Cry1Ab) suffered greater mortality than control lacewings fed on non-Bt maize-fed prey. Only 37% of the C. carnea fed Bt maize-fed larvae of the cotton leafroller, Spodoptera littoralis, or the European corn borer, Ostrinia nubilalis, survived, whereas 62% of the control group fed on non-Bt maize-fed caterpillars survived. In a subsequent study, using an artificial liquid diet it was determined that immature C. carnea were sensitive to the Cry1Ab toxin at a level of 100 μg/ml of diet.

However, the level of Cry1Ab in maize is about 4 μg/g fresh weight, which is considerably less than 100 μg/ml. At present there appears to be no explanation for the discrepancy between the artifical diet and exposure via prey. The apparent sensitivity of C. carnea to the Cry1Ab is an interesting labouratory observation, and should be validated with histological and receptor-binding studies.

The most widely publicized study of the potential impact of Bt maize on a non-target insect is the finding that larvae of the monarch butterfly, Danaus plexippus, are sensitive to certain doses of Bt maize pollen. In this labouratory study, milkweed leaves were covered with Bt-maize pollen and then fed to larvae. Control larvae were fed on milkweed leaves covered with non-Bt pollen or untreated milkweed. The key finding of the study was that the larvae fed milkweed leaves treated with unquantified amounts of Bt pollen had a lower survival rate (56%) than controls (100%).

The authors stated that their results had 'potentially profound implications for the conservation of monarch butterflies' because the central corn belt where Bt maize adoption by farmers continues to grow is also an important habitat for monarchs. In assessing the relevance of these findings on the green lacewing and monarch larvae, or other non-target organisms for that matter, it should be kept in mind that bacterial insecticides based on Bt should be just as toxic if not more so because they contain more Cry proteins than Bt crops.

In other words, monarch larvae that feed under field conditions on milkweed leaves treated with a product that contains B. thuringiensis subsp. kurstaki, from which the Cry1Ab protein gene was derived, will be equally if not more sensitive to the bacterial insecticide. Similarly, C. carnea immatures that feed on caterpillars intoxicated as a result of feeding on a Bt insecticide will be equally sensitive to the activated toxins in these larvae. So the issue here is not so much one of Bt crops, but whether Cry proteins will impact beneficial insects regardless of the source.

Despite the sensationalistic attention given to the monarch study by the popular and even the scientific press, the design of this study has been criticized by scientists familiar with monarch biology and maize production. The principal criticisms are that monarch larvae were not fed well-quantified amounts of Bt maize pollen, that larvae given a choice may find milkweed contaminated with Bt pollen unpalatable and thus should have been given a choice, that Bt pollen is only present for 7-10 days of the growing season, and that the most Bt pollen dispersed remains within several meters of the dispersal in the field. Thus, maximum monarch exposure would normally be limited to larvae developing within or at the edges of corn fields, but not in other habitats.

As a result, only monarch larvae developing in very close to Bt corn fields during pollen-shed would typically encounter quantities of Bt pollen grains on the surface of milkweed leaves. Extraordinary attention was given to the preliminary findings in the scientific and popular press on the potential negative effects of Bt pollen on monarch populations.

A benefit of this attention was that it resulted in a series of collabourative studies in 1999 and 2000 devoted to a much more rigorous assessment of these potential negative effects under field conditions throughout the US corn belt and Canada, as well as in the labouratory. The overall conclusion of these studies is that the effects of Bt maize on monarch populations will be negligible, especially in comparison with the effects of using chemical insecticides to control maize pests.

The basis for this conclusion was that most of the registered Bt maize varieties produced only a low level of Bt protein, and even moderately high levels of consumption of this pollen by larvae did not affect their development or survival. Other 'mitigating' factors that limited exposure of larvae to Bt proteins in pollen included limited overlap between larval feeding periods thoughout the corn belt and the period of pollen shed, the relatively limited adoption of Bt maize (19%), and the widespread availability of milkweed outside areas where maize is grown.

Significantly, larvae in maize plots treated with the chemical insecticide (-cyhalothrin) showed no weight gain because all were killed by the insecticide within 24 hours of treatment. The experimental design used to assess risk in these studies provides a good model for other studies in that the critical aspects of the assessment focused on experiments under field conditions, further supported by labouratory experiments.

Moreover, the relative risks to a non-target organism of using Bt maize or a chemical insecticide were compared, thereby eliminating the 'double-standard' that is often applied to assessing risks of Bt crops, in other words, applying standards to Bt crops that are much higher than those applied to chemical insecticides. Regardless of whether the results obtained against non-target organisms in labouratory studies show favourable, unfavourable or neutral effects, these must be followed by long-term studies under field conditions. The reason is that labouratory studies are designed to reveal any potential adverse effects by exposing non-target organisms to excessively high levels of Bt proteins, levels that would not be encountered under field conditions. Moreover, field studies should include comparisons to current agricultural practices, which often will include the use of chemical insecticides.

At present, there have only been a few studies so far that have evaluated the effects of Bt crops on non-target organisms under field conditions over the length of the growing season. These studies are nevertheless important because non-target organisms and their prey were exposed to Bt Cry proteins in the form synthesized in the crop and over a continuous period at an operational level. The non-target organisms studied under field conditions have all been insects and the test crops either Bt maize or Bt cotton. The insects consisted of a plant bug and four beneficial insects, specifically two parasites and two predators, one of which was the lacewing, C. carnea.

In these season-long studies, no adverse effects were observed on any of the non-targets under field conditions, even on C. carnea, is apparently sensitive to Cry1Ab. Other non-target organisms that have recently received attention with respect to the environmental effects of Bt crops are soil microorganisms.

Some in the labouratory, others under quasi-field conditions, it has been shown that Bt proteins such as Cry1Ab released into soil can bind to various types of soil particles and remain active for periods of at least several months. The consequences of this accumulation remain unknown, but recent studies by Stotzky and his colleagues indicate there were no significant effects on soil microorganisms or earthworms.

Evidence from other studies indicates that Bt Cry proteins are rapidly degraded after incorporation of Bt crop tissue into soil, although low levels can persist for at least several months. While these studies indicate that the temporary accumulation of Bt Cry proteins in soil have no adverse effects, longer term studies are needed to further evaluate these initial results.

## SAFETY OF BT CROPS TO NON-MAMMALIAN VERTEBRATES

With respect to non-mammalian vertebrates, tests of Bt maize have been carried out using bobwhite quail, channel catfish and broiler chickens, the latter in a 38-day study in which the chickens were fed on Bt maize grain. In the chicken studies, chickens fed Bt maize grain were compared with chickens fed non-Bt grain. The variables examined were survival, weight gain and feed efficiency. No significant differences were found between the chickens fed Bt maize grain versus non-Bt maize grain. Similar results were obtained in other studies in which the effects of feeding Bt maize versus non-Bt maize to chickens were studied.

## Safety of Crops to Mammals

There have been very few studies published in the scientific literature in which mammals were fed on Bt crops and then monitored for effects. Most of the safety data used to register Bt crops, is based on studies of Cry proteins tested as bacterial insecticides or produced in E. coli. But there has been at least one comprehensive study in which the effect of feeding rats on Cry1Ab tomatoes for 91 days was examined, and amplified with other studies using Cry1Ab produced in E. coli. In this study, weanling Wistar rats were fed for 91 days on a lyophilized powder of Cry1Ab tomatoes (40.6 ng Cry1Ab/mg protein) that made up 10% of the diet.

Control rats were fed a similar diet consisting of a 10% non-Bt lyophilized tomato powder. In terms of human consumption, this amount was calculated to be the equivalent of a human eating 13 kg of tomatoes per day. No significant differences were observed between the treated and control rats in terms of survival, food intake, body and organ weights, or other gross macroscopic examinations of tissues.

The lack of any apparent adverse effects resulting from feeding rats on a high dose of Cry1Ab tomato corresponds to results obtained in feeding rats bacterial toxins containing Cry1Ab. In related studies, these investigators used Cry1Ab produced in E. coli to study the effects and fate of this protein on various mammals and mammalian tissues.

Histological binding assays were conducted to determine whether the Cry1Ab could bind to the intestinal mucosa of mice, rats, rhesus monkeys and humans. Cry1Ab did not bind to microvilli in these mammals in any of the intestinal regions, including the esophagus, stomach, duodenum, jejunum, ileum and colon, though the toxin did bind to microvilli along the entire length of the larval midgut in the tobacco hormworm, Manduca sexta, the lepidopteran used as the insect control. In a separate study, gastrointestinal tissues were examined 7 hours after rats were fed an amount of Cry1Ab equivalent to a human consuming 2000 kg of tomatoes in a single day.

No binding was detected in any of the above intestinal regions. These results indicate that none of the mammalian species tested had receptors for Cry1Ab on their intestines. In other studies, it was found that the active toxin form of Cry1Ab (66-68 kDa) was degraded to peptides of 15 kDa or less within 2 hours when incubated in simulated human gastric juices at pH 2. In studies using Cry1Ab consumed ad libidum in drinking water for 31 days by New Zealand white rabbits in an amount equivalent to humans consuming 60 kg of tomatoes per day, no differences were found between treated and control groups.

Histological studies of the stomach and intestines showed no abnormalities, and no antibodies were induced against Cry1Ab in the treated rabbits. In another study of a Bt crop, Bt potato producing an unidentified Cry1 protein was fed to one-month-old mice for two weeks, after which the mice were sacrificed and sections of the ileum were examined by light and electron microscopy.

The control mice were fed on non-Bt potato. No statistically significant differences were observed between the mice fed Bt potato and the control mice. In another Bt potato study, raw potatoes containing the Cry3A protein (used to control the Colourado potato beetle) were fed to rats for 28 days. The control was the same potato variety that did not produce Cry3A. The rats were fed levels of Cry3A corresponding to a human consuming several kilograms of potato per day for 28 days. No adverse effects were observed.

The treated and control rat groups were similar in weight gain and behaviour. In other studies on the effects of Bt grain on the health and perfomance of other animals, no significant differences were found between groups fed Bt versus non-Bt grains in dairy cows, beef cows or steers. This combination of studies using a variety of methods and feeding studies that include Bt crops and candidate Cry proteins validate the use of these methods for assessing safety and risk, and support the use of data developed for the registration of bacterial insecticides to support the safety of Bt crops.

Most studies of Bt crop safety have focused on tests designed to detect potential harmful effects. There have been few studies of potential improvements in safety, though it is often noted that Bt crops should be safer because they reduce chemical insecticide usage. Yet one notable improvement in safety that has not received much attention is that Bt maize grain contains lower levels of certain fungal toxins known as fumonisins. These toxins can cause illness and death in horses and pigs, and have been implicated in certain forms of liver and esophageal cancer in humans.

Fumonisins result from ear rot infections of maize by Fusarium species, which invade tissues damaged by the European corn borer, O. nubilalis. In Bt maize grain, the levels of fumonisins are reduced by as much as 93% in comparison with non-Bt maize grown under similar conditions. Future of Bt crop safety in perspective Insecticides based on endotoxin proteins of B. thuringiensis have been in use for 40 years, and have a safety record for non-target invertebrates and vertebrates including mammals that far surpasses that of any synthetic chemical insecticide.

This safety record, combined with the efficacy of certain Bt Cry proteins and the advent of recombinant DNA technology, led to the development of transgenic insect-protected Bt crops that are being adopted rapidly, especially Bt cotton and Bt maize, by farmers in the United States and a few other countries. The safety of these crops for non-target organisms is based primarily on data from numerous toxicological studies carried out on bacterial isolates used to register Bt insecticides, and to some extent on Cry proteins produced in E. coli or Bacillus species.

There have been comparatively few studies so far that have directly evaluated Bt crops in feeding studies. Those that have been done provide strong evidence that Bt crops are safe for humans and other animals as well as for most non-target invertebrates, with the exception of insect species closely related to target insect pests, and a few other species. There is some evidence based on forced-feeding labouratory studies that non-target organisms such as larvae of the monarch butterfly and lacewings may be adversely affected or killed by Bt proteins produced in Bt crops.

But the limited number of field trials conducted to date have found no evidence that Bt crops adversely affect generalist insect predator populations or parasite populations. In fact, evidence is mounting that Bt crops may enhance biological control and integrated pest management programmes by reducing the amounts of synthetic chemical insecticides used to control insect pests, thereby maintaining the populations of beneficial insects, which are typically decimated by the use of chemical insecticides.

In addition, Bt crops such as Bt maize may actually be safer for consumption by mammals than non-Bt crops because the control of lepidopteran pests reduces the level of certain mycotoxins that result from fungal infections following insect damage. Whereas existing evidence attests to a high degree of safety for Bt crops, these should be more extensively evaluated directly through detailed feeding and toxicological studies, and through more extensive and long-term ecological studies to determine their effects on non-target invertebrate populations under operational field conditions.

Given the public controversy that has arisen over genetically engineered crops, numerous such studies are underway, and will find their way into the refereed scientific literature in the coming years. In designing safety studies for non-target mammals, more emphasis should be placed on feeding studies in which animals are fed Bt crops, rather than relying solely on surrogate data developed from feeding studies using Bt insecticides or Cry proteins produced in E. coli. This should at least be done for a range of Cry proteins used or destined for use in a variety of Bt crops to allay the concerns being raised by public.

The costs will be minor compared to the investments being made in Bt crops and the advantages that will accrue long term from having high public confidence in the safety of transgenic crops. Though there are technical hurdles that must be overcome in 'direct' feeding studies due to the low levels of Cry proteins in Bt crops, tests like those developed by Noteborn *et al.* indicate that such studies can be done. The results of less direct methods, such as chemical fingerprinting of metabolic changes that might be detected between Bt crops and their isogenic parents, would appear unnecessary and could easily lead to questionable interpretations.

For example, what if the metabolic pool of a mRNA for a certain enzyme is found to be twice as high in a Bt tomato as opposed to its isogenic parent, but all the direct feeding studies show that the tomato is safe? And what if the mRNA for the same enzyme in another tomato variety is shown to be just as high as the mRNA in the Bt tomato. Do any of these results mean the Bt tomato is not safe? It is important that any new tests developed to evaluate safety be directly relevant to providing data that can serve as a scientific basis for the further assessment of safety.

Of course, all methods used need to be subjected to rigorous validation. Studies that result in data of questionable value or that could be subject to misinterpretation should be avoided. Aside from the scientific rationale for taking this approach, a reason for doing this is that many individuals and environmental groups are opposed to Bt crops simply because they object to the use of genetic engineering technology to modify crops. A good example of this sort of thinking comes from Lord Peter Melchett, a former Labour Minister in the United Kingdom, and a leader in Greenpeace's effort to stop the use of biotechnology in agriculture.

Lord Melchett opposes the use of science to make decisions on food consumption, preferring that selections be made on an emotional basis, stating that 'If its acceptable to choose our car based on emotion and not science, why should it be wrong to choose our food that way?' The public cannot afford to let this sort of thinking dominate or influence views and decisions on what is safe to eat. Therefore, to counter such thinking and avoid the misuse of data, meaningful short- and long-term studies like those carried out by Noteborn *et al.* should be the type used to evaluate Bt crop safety.

Such studies should detect any adverse effects of natural or modified proteins synthesized by crop plants. In addition, the same rationale is relevant to justifying long-term, i.e. studies of several years, on the effects of Bt crops on non-target organisms, including soil microorganisms to assess any potential ecological effects, be they positive, neutral or negative. The results of initial studies indicate that Bt crops will be much safer for non-target organisms, especially beneficial insects, than chemical insecticides.

If long-term studies confirm these initial results, dissemination of this knowledge should improve public confidence in this new and important crop protection technology. In summary, Bt crops are one of the first practical examples of new and powerful pest-control technologies emerging from the results of decades of basic research and recombinant DNA technology. An overwhelming amount of evidence based on studies of Bt Cry proteins indicates these crops are safe for consumption by humans and other vertebrates, and are much safer for non-target invertebrates and the environment than synthetic chemical insecticides.

Based on the scientific evidence supporting these conclusions, the US Environmental Protection Agency will continue to register Bt crops for use in the United States. While additional direct testing of Bt crops is warranted to further assess their safety, there is no reason at present to think that these crops present risks greater than those associated with the consumption of non-Bt crops. In fact, Bt crops may be safer for human consumption than conventional crops because they contain lower levels of mycotoxins and residues of chemical insecticides.

## Chapter 7

# Microbial Pesticidal Agents

Baculoviruses are a group of viral pathogens that cause fatal disease in arthropods. More specifically, these baculoviruses infect primarily only members of the class Insecta and even within this class limit themselves to only a few orders, i.e. Lepidoptera, Diptera, Hymenoptera and Coleoptera. Baculoviruses have been studied extensively dating back to the 1960s. The literature is filled with a plethora of information on their biology and safety. The wild-type baculoviruses have been evaluated for use as insecticides numerous times, especially during the late 1960s and up into the early 1970s. The first baculoviral product introduced in the commercial arena was Elcar. Three other non-commercial preparations produced and used by the US Forest Service for control of forestry pests were developed and registered under the names of Gyp-Chek (Lymantria dispar NPV, LdMNPV), EM Biocontrol-1 (Orgyia pseudotsugata NPV, OpMNPV) and NeoCheck-S (Neodiprion sertifer NPV, NsMNPV).

These products have received only limited use. However, recently the Elcar product was reborn under the trade name of GemStar™ and is being sold in various countries around the world. Intricate evolutionary relationships have developed between the baculoviruses and the arthropods that they exploit. This close evolutionary relationship has resulted in the restriction of host range. Over the past 40 years, extensive studies evaluating the safety of baculoviruses against vertebrate species have been carried out.

A summary of the data indicates that over 26 different insect baculoviruses have been tested for pathogenicity against 10 different mammalian species, including rats, dogs, mice, guinea-pigs, monkeys and humans. The methods of administration in these tests varied, including oral, intravenous injection, intercerebral injection, intramuscular injection and topical application. In no case was there any indication of toxicity or allergenicity. These results have been summarized in a number of reviews. Specificity is also a major factor in making baculoviruses safe to non-target insects. Baculoviruses generally only infect insects in a few families. In some cases only one species in a genus may be susceptible (e.g. LdMNPV). Those insect species that are considered beneficial (control pest insect species) are generally in families other than those targeted by baculoviruses. There are a few exceptions to this rule, however, as they can also infect some desirable lepidopterans.

This insect specificity makes baculoviruses excellent candidates for use in integrated pest management (IPM) programmes as they are compatible with other insect control technologies (bioinsecticides, chemical insecticides and transgenic plants), and have no effect on any non-target insects such as honeybees. If baculoviruses have all of these attractive parameters, why are they not widely used as insecticides? Baculoviruses are unable to rapidly kill a susceptible insect.

It may take 5-15 days to kill a host larva following infection. This period is affected by such factors as temperature, dosage and larval age. Younger larvae are generally more susceptible to infection and die sooner, but even though they are infected they continue to eat, usually up until the day they die. Obviously a crop with a low economic threshold cannot tolerate this continued feeding damage until the insect dies. Growers want to see dead insects soon after the application of any insecticide.

Fortunately, recombinant technology offers a new life to the old concept of utilizing baculoviruses as insecticides. It is now possible to genetically engineer a baculovirus with a toxin gene, giving it improved insecticidal properties and increasing the speed of action. These toxin genes usually code for an insect-specific, proteinaceous toxin. The genes can be isolated from the venom of certain arthropods (e.g. scorpions or spiders).

Typically, when a baculovirus containing a toxin gene infects an insect, the feeding is rapidly halted and causes a quick death. Insertion of the insecticidal gene does not affect the LD50 of the baculovirus but does significantly hasten the killing speed through the pharmokinetic interaction of the toxin within the insect. Even before actual death, the toxin causes cessation of feeding resulting in less plant damage, then paralysis, and eventually death.

Diet-overlay assays using an LD99 quantity of recombinant NPV have shown that, when second-instars of susceptible insect species are exposed to the wild-type virus versus the genetically engineered virus, at 10 days the LD50 is the same for both groups; however, the wild-type infected larvae characteristically take as long as 5.1 days to die versus 2.6 days for larvae infected with genetically engineered virus. Greenhouse studies have shown that this speed of kill translates to significantly better plant protection. In one greenhouse study, cotton plants sprayed with a genetically engineered virus containing a scorpion toxin gene were compared with plants treated with a chemical standard, esfenvalerate. The plants were then artificially infested with target insect species and then evaluated 4-5 days after treatment. The cotton incurred 18.9% versus 13.0% damaged squares for the two treatments, respectively. Statistically, both treatments were equal. Numerous US field trials have been conducted, again showing that the speed of kill is influential in reducing the damage.

One study conducted in North Carolina (USA) on tobacco against the tobacco budworm, showed that the genetically engineered virus performed as well against this insect as did the standard, acephate. Additionally, toxin production is limited only to the body of the insect (i.e. there is no release of toxicant into the environment as with conventional insecticides).

This makes the genetically engineered baculovirus a prime candidate for insecticidal use, and in some situations makes it competitive with chemical insecticides. Block and Zhihong have given a good summary of the progression of the technology in engineering of baculoviruses, culminating with the actual field release of a toxin gene carrying baculovirus by a commercial company in 1995.

## PUBLIC RELATIONS ACTIVITY

These genetically manipulated baculoviruses obviously need approval by regulatory agencies prior to any field release and eventual registration. Using the scientific process to demonstrate safety of a genetically engineered organism to the public is only part of the 'proof of safety' process. This information has to be transmitted to the public to establish the true level of safety in their perception. Biotechnology has come under public scrutiny to varying degrees depending on where one is in the world. Environmental activists have attacked biotechnology from scientific, ethical and personal viewpoints. These activities have created doubts, which must be addressed, no matter how trivial or unscientific they may appear. Obviously the development of a plan is necessary.

Assuming that approval of regulatory agencies is adequate (to make a release) could be a major mistake, as demonstrated by the experience of the National Environment Research Council (NERC) Institute of Virology, Oxford, England. All the necessary regulatory clearances were obtained for a release, but no effort had been made to inform the public around the application site. The end result was that the public felt that 'a fast one was being pulled'. The researchers then had to backtrack and attempt to educate an irate public. American Cyanamid Company conducted field releases in 1995, 1996 and 1998 of baculoviruses containing an insect-specific toxin gene from either the scorpion Androctonus australis or the straw-itch mite Pyemotes tritici. Before any release was attempted, American Cyanamid Company scientists developed a safety database (extracted from the literature) and through carefully conducted experimentation.

This database was summarized and written into a brochure. Also, a video was produced, which explained the engineering process and the resultant recombinant baculovirus' impact on the susceptible host. Both the brochure and video were made available for public distribution. This information was utilized at numerous open meetings with representatives from many areas, academia, regulatory, mass media, environmental groups, politicians and the general public. Thus communication lines were opened and the dialog that developed served to mitigate any concerns. Early on, American Cyanamid Company scientists opened communication lines with responsible national environmental groups to explain the technology.

Gaining the support of these groups inevitably helped ease any concerns that may have been present. Using the basic elements of risk communication, a policy was adopted that all the safety information available would be made public. This policy was followed throughout the public education process and in all US EPA submissions (i.e. all safety data supplied to the US EPA were available to anyone for the asking). This strategy resulted in the generation of very little negative response from the aforementioned groups.

## RISK ASSESSMENT

In addition to evaluating the insecticidal properties of baculoviruses against targeted pest species, it is essential also to assess the potential environmental impact of such gene-inserted pathogens prior to, as well as during, initial small-scale field trials. In their review of ecological considerations pertaining to recombinant baculoviruses, Richards *et al.* noted that an environmental impact evaluation must address the effect of the recombinant entity on non-target species and populations, as well as the impact of the genetic modification on the ability of the virus to persist in the environment, relative to that of the wild-type or feral counterpart.

It is important to determine whether the addition of a foreign gene alters the host-range of the virus, and whether the insecticidal protein produced in the virus-infected insect will present a hazard to other species. Additionally, comparative testing of recombinant and feral baculoviruses for physical characteristics, genetic fitness and dispersal properties can provide insight as to relative stability in the environment.

## Recombinant Nucleopolyhedroviruses

Nucleopolyhedroviruses isolated from alfalfa looper, Autographa californica (AcMNPV), and cotton bollworm, Helicoverpa zea (HzNPV), have been engineered as vectors for carrying and expressing insecticidal genes into the body cavities of pestiferous insect species. AcMNPV is known to infect about 39 species of Lepidoptera. Larvae of tobacco budworm, Heliothis virescens (F.), and cabbage looper, Trichoplusia ni (Hubner), are among the most permissive to infection by AcMNPV, whereas others such as H. zea are only moderately sensitive, or semipermissive, to this virus.

Conversely, HzNPV is highly virulent against larvae of the entire Heliothine spp. complex, i.e. H. virescens and Helicoverpa spp. A number of insect-selective toxin genes have been inserted into the genomes of AcMNPV and HzNPV in an attempt to make them more effective as biocontrol agents in cropping systems. Two such genes, AaIT and LqhIT, which were isolated from venoms of the Algerian scorpion, Androctonus australis (Hector), and Israeli yellow scorpion, Leiurus quinquestriatus hebraeus Birula, respectively, are known to inappropriately modulate neuronal sodium channels.

The protein encoded by AaIT acts as an excitatory sodium channel agonist, causing repetitive firing of the insect's motor nerves and overstimulation of skeletal muscle. Symptoms exhibited by insects intoxicated by AaIT include cessation of feeding, paralysis and death. Conversely, LqhIT is a depressant neurotoxin, causing a blockade to sodium conductance and suppression of axonal depolarization and neuronal action potentials. LqhIT blocks neuromuscular transmission and induces progressive paralysis in affected insects.

A third invertebrate toxin gene which has been incorporated into selected baculoviral genomes is tox34, which was cloned from the genomic array encoding venom of the straw-itch mite, Pyemotes tritici. In assays conducted on primary cultures of motor neurons isolated from selected lepidopteran species, it has been shown that the product of the tox34 gene, a 33-kDa protein designated as TxP-I, causes a blockade of presynaptic inward calcium currents, which likely results in reduced secretion of the neurotransmitter acetylcholine.

The recombinant baculoviruses AcMNPV-AaIT, AcMNPV-LqhIT, AcMNPV-tox34, HzNPV-AaIT and HzNPV-LqhIT have been evaluated for insecticidal efficacy in open field trial settings. Considerable research has been devoted to assessing mammalian and environmental safety of these aforementioned baculoviruses.

## Selectivity of Isolated Toxins

Studies conducted to define the pharmacology of purified AaIT and LqhIT have shown these toxins to possess remarkable specificity. Although both of these toxins are very potent against insects, they appear to have no effect on mammals even at very high dosages. Using hemocoelic injection methodology, DeDianous *et al.* determined that LD50 values for AaIT against selected species of Diptera and Orthoptera ranged from 20 to 3770 ng/kg.

Conversely, these same researchers found that subcutaneous injection of AaIT, at dosages as high as 50 g/kg, caused no adverse effects in mice. In separate oral and inhalation dosing studies conducted on labouratory mice, Possee *et al.* showed that AaIT at 1 μg/animal had no effect on the animals' body weight, dietary intake, survival or appearance of internal organs at necropsy.

Experiments in vitro and in situ have demonstrated that AaIT causes convulsive firing and overstimulation of insect motor neurons and skeletal muscle, whereas it caused no effects in neuromuscular preparations isolated from mammals and selected species of Crustacea and Arachnida. Further, in experiments conducted on isolated nervous tissue, radiolabeled AaIT readily bound to such tissue from insects but not to nervous tissue from crustaceans or mammals.

LqhIT has also demonstrated a high affinity for insect neuronal receptors and a high degree of potency against species within this class, but no detectable toxicity to mammalian or crustacean species. Researchers at DuPont Agricultural Products intravenously injected individual mice with LqhIT at a dosage of 0.5 mg/kg and then observed them for 14 days. None of the mice exhibited any clinical signs of toxicity over the duration of the study, and necropsies performed on the animals at the conclusion of the test revealed no gross lesions on tissues and organs as a result of exposure to LqhIT.

Compared with AaIT and LqhIT, studies pertaining to pharmacology of TxP-I are limited. However, Tomalski *et al.* reported that the PD50 (paralysis dosage) of this straw-itch mite toxin in larvae of wax moth, Galleria mellonella, was about 0.5 mg/kg, whereas TxP-I had no effect on labouratory mice at dosages as high as 50 mg/kg.

## Vertebrate Safety of Recombinant NPVs

It is intuitive to expect that recombinant NPVs would have no impact on mammals, birds or fish, because the NPVs would have to infect cells of these vertebrate species in order to produce toxin. It is well documented in the scientific literature that NPVs are highly specific for insect hosts, with AcMNPV and HzNPV able to infect only certain species of Lepidoptera. However, it has been considered prudent to repeat such vertebrate pathogenicity/toxicity studies with the recombinant forms of NPVs.

Possee *et al.* conducted a series of studies with AcMNPV-AaIT to determine its acute oral, dermal and subcutaneous toxicity or pathogenicity to rats and guinea-pigs. Over a 14-day period following a single dosing of each animal with 1 × 106 occlusion bodies (OBs) of recombinant NPV, all rats and guinea-pigs remained healthy and showed no abnormalities in weight gain, dietary intake, physical appearance or motor skills.

Additionally, necropsies performed at the end of the study showed that no gross changes occurred in organs of treated animals. When used as a bioinsecticide, a recombinant NPV would be formulated and sprayed onto crops in the form of occlusion bodies. Since toxin would not be present in OBs of a recombinant NPV, the only route of exposure to the toxin by vertebrates would be through ingestion of infected insect larvae or herbivory of foliage contaminated with insect cadavers.

In an attempt to mimic this route of exposure, researchers at American Cyanamid Company fed labouratory mice a diet which consisted partly of H. virescens larvae that had been previously infected with AcMNPV-tox34. Prior to use in the study, H. virescens were exposed to 10 times the larval LD95 of AcMNPV-tox34. Three days after exposure, larvae that exhibited paralysis (i.e. indicative of TxP-I production and intoxication) were homogenized and fed to mice by gavage.

Each mouse was fed the equivalent of four infected H. virescens larvae on a daily basis for five consecutive days. No clinical signs of toxicity or effects on survival were seen in treated mice during the study period. Dietary intake and body weight gains for treated and non-treated mice were equal throughout the five-day study. Macroscopic evaluation of tissues and organs in mice which were sacrificed at the end of the five days of dosing revealed no treatment-related pathological changes or abnormalities.

American Cyanamid Company evaluated AcMNPV-AaIT in a series of studies which were established by the US EPA to ascertain the safety of microbial pesticides. In an acute oral toxicity-limit test, AcMNPV-AaIT had no impact on survival of labouratory rats, nor did it impact internal organs of the animals. The budded form of the recombinant virus was also evaluated for its ability to infect mammalian cells (i.e. diploid human lung cell line MRC-5). AcMNPV-AaIT did not alter cell doubling time or cell morphology, and as expected, there was no evidence of viral replication or expression of AaIT. Mallard ducks, bobwhite quail and rainbow trout were fed diets consisting of AcMNPV-AaIT OBs as well as H. virescens larvae infected with AcMNPV-AaIT, and no adverse effects were observed in any of these three vertebrate species. Finally, no abnormalities were observed in the aquatic invertebrate, Daphnia magna, when it was maintained in aqueous suspensions of AcMNPV-AaIT. Researchers at DuPont Agricultural Products compared feral AcMNPV and AcMNPV-LqhIT for infectious and pathogenic behaviour in labouratory rats using oral, intravenous and inhalation routes of exposure.

Neither virus was infectious or toxic to rats by any of the aforementioned means of exposure at dosages ranging from 1 × 107 to 1 × 108 OB/animal. DuPont researchers also confirmed that AcMNPV-LqhIT was not infective or cytotoxic to human liver (Chang liver), intestine (strain 407) and lung cell lines. Additional studies showed that bobwhite quail receiving oral and intraperitoneal exposure to OBs, and trout and grass shrimp exposed to the NPV via food and/or aquatic environment, were unaffected by AcMNPV-LqhIT.

### Invertebrate Selectivity of Recombinant NPVs

In order for toxin to be expressed by a recombinant baculovirus, the virus must first establish a productive systemic infection within the host (particularly if the toxin gene is under transcriptional control of a late promoter). If an organism is non-permissive to infection by the vectoring agent, the foreign gene will not be transcribed, i.e. host-range is determined by viral gene products rather than by those of the introduced gene.

Using either hemocoelic injection or a per os inoculation methodology, Huang *et al.* tested budded and pre-occluded forms of the following viruses for their abilities to establish an infection and replicate within selected non-lepidopteran insects: AcMNPV, LdMNPV, OpMNPV and Bombyx mori NPV (BmNPV). Each of the aforementioned viruses was engineered to express an inserted reporter gene (e.g. luciferase, α-galactosidase), so that even symptomless infections could be identified within the insects (i.e. viral replication could be based on detection of reporter gene product).

Huang *et al.* demonstrated that individuals from 13 species of insects, representing Blattodea, Coleoptera, Diptera, Hemiptera, Homoptera, Neuroptera and Orthoptera, did not support detectable replication of any of the four viruses included in the study. Maximum challenge labouratory studies have been conducted to determine the impact of certain recombinant NPVs on growth, behaviour and survival of numerous species of non-lepidopteran invertebrates. Depending on species or growth stage tested in these labouratory assays, animals were exposed to AcMNPV-AaIT, AcMNPV-LqhIT, AcMNPV-tox34 and/or HzNPV-AaIT via ingestion of contaminated artificial diet or foliage (OBs), consumption of virus-infected prey.

Results from these labouratory studies showed that none of the aforementioned recombinant NPVs directly caused adverse effects in individuals of the following predatory, parasitic, herbivorous or saprophytic invertebrate species: green lacewing (Chrysopa carnea Stephen), insidious flower bug (Orius insidiosus (Say)), ground beetle (Pterostichus modidus), honeybee (Apis mellifera), hymenopterous parasitoid (Microplitis croceipes), twospotted spider mite (Tetranychus urticae Koch), boll weevil (Anthonomus grandis grandis Boheman), western corn rootworm (Diabrotica virgifera virgifera LeConte), Japanese beetle (Popillia japonica Newman), earthworm (Lumbricus spp.), Chinese mantid (Tenodera aridfolia), funnel web spider (Ixeuticus spp.), subterranean termite (Reticulotermes flavipes Kollar), convergent lady beetle (Hippodamia convergens), bigeyed bug (Geocoris spp.), red imported fire ant (Solenopsis invicta Buren), the social wasp (Polistes metricus (Say)), German cockroach (Blatella germanica) and common fruitfly (Drosophila melanogaster).

In a series of small-scale field trials conducted on cotton, tobacco and cabbage, it was determined that weekly foliar sprays of AcMNPV-AaIT, at rates as high as 2.0 × 1012 OB/ha, had no effects on diversity or density of non-target arthropods, relative to arthropod populations in non-treated plots of each crop. At least 45 species, representing 26 families of insects, spiders and mites, were found in collections taken from these small-plot field tests. Heinz *et al.* found that plots of field-grown cotton which were sprayed with either HzNPV-LqhIT or AcMNPV-LqhIT had similar populations of non-target predatory arthropods as those plots treated with feral forms of the viruses.

Further, cotton treated with the aforementioned LqhIT-inserted viruses had significantly greater numbers of beneficial arthropods than plots treated with conventional insecticides, such as the pyrethroid esfenvalerate and the semi-synthetic macrolide emamectin benzoate. Published and non-published studies have also shown that even within Lepidoptera, recombinant NPVs exhibit a host-range similar to that of their feral counterparts. For example, in a series of dose-response assays conducted on larvae of 52 different species of Lepidoptera, Bishop *et al.* found that AcMNPV-AaIT exhibited potency or virulence against each species at a level similar to that imparted by feral AcMNPV. There have been results from time-limited assays which would appear to suggest that AcMNPV-semipermissive H. zea might be more sensitive to gene-inserted forms of AcMNPV than the feral form of this virus. However, studies such as those reported by Treacy and All were designed to compare recombinant and feral viruses for speed of insecticidal activity against lepidopteran pest species, and thus focused on levels of larval mortality during the initial 8-10 days of viral exposure.

Although such studies can demonstrate potential of a recombinant virus to be an effective insecticide, they do not provide insight as to the overall impact of such a gene-inserted NPV, relative to is feral counterpart, on the life cycle of a semipermissive lepidopteran species. Therefore, entomologists at American Cyanamid's Agricultural Research Centre in Princeton, New Jersey recently conducted a labouratory study to compare wild-type AcMNPV, AcMNPV-AaIT and AcMNPV-tox34 for effects on survival of AcMNPV-semipermissive H. zea at selected points in time throughout its life cycle, i.e. middle and late larval development and adult eclosion (or 5, 10 and 35 days after placing second-instars on viral contaminated diet).

Based on temporal LD50 values and presence of non-overlapping fiducial limits, the two aforementioned recombinant AcMNPVs were about 800 and 25 times more lethal to H. zea than feral AcMNPV at 5 and 10 days after initial exposure to the viruses, respectively. However, when LD50 values were calculated from final test-population responses to treatments (i.e. 35 days after test initiation), it was found that AcMNPV-AaIT, AcMNPV-tox34 and AcMNPV were statistically equal in virulence against H. zea.

Results from this later study indicated that when the feral and recombinant forms of AcMNPV were allowed to complete their infection cycles in AcMNPV-semipermissive H. zea, there were little to no difference among the NPVs in virulence. The differential H. zea LD50 values between the two recombinant viruses and wild-type AcMNPV seen during the first 10 days for the study were likely due to pharmacology of the expressed toxins (per the recombinant NPVs), and not a result of any genome-related changes in viral infectivity.

### Environmental Stability of feral and Recombinant NPV

The potential for recombinant NPVs to displace their feral counterparts in the environment would be dependent upon expression of an advantage in fitness by the genetically altered virus. Fitness of a virus is determined by its physical characteristics and capacity for production of active progeny. Evidence gathered to date indicates that NPVs which have been engineered for enhanced insecticidal speed without significant alteration in host-range are likely to be less competitive in the environment than corresponding feral viruses.

Insertion of a gene encoding an insecticidal protein does not impact the physical or biochemical properties of viral occlusion bodies. Thus, such recombinant and feral NPVs are equally sensitive to abiotic and biotic degradation factors. Photolysis and photo-mediated oxidation are the primary mechanisms for inactivation of baculoviruses. It is thought that ultraviolet (UV) sunlight in the UV-B spectrum can directly cause strand breakage in viral DNA, or lead to production of highly reactive radicals (e.g. peroxides) which, in turn, degrade viral particles. Field studies have shown that selected feral NPVs which were applied to cotton foliage lost 70-80% of their initial biological activity in less than four days.

Similarly, following spray-application of a wettable powder formulation of AcMNPV onto field-grown cotton, American Cyanamid Company researchers found that 3-hour-old foliar residues of the virus caused 98% mortality in populations of H. virescens larvae, whereas larval mortality was only 5% on cotton leaves that had been aged in the field for two days. Although NPVs are rapidly inactivated when exposed to sunlight, their persistence when buried in soil can be for months or years.

In their review of viral persistence in soil, England *et al.* highlighted the following as factors that can impact stability of viral OBs and DNA:

- Proteolytic microorganisms (e.g. bacterium);
- Soil pH;
- Soil temperature (e.g. freezing and thawing);
- Soil-binding capacity;
- Soil moisture;
- Predation by soil-dwelling protozoa.

A characteristic that limits the survivability of recombinant NPVs is the reduced ability to produce viable progeny. The yield of progeny occlusion bodies from a baculovirus infection is a key factor controlling viral population dynamics in the field. The slow killing speed imparted by a feral NPV on its host is strategic, because the virus can continue to replicate and maximize the progeny numbers while the infected larva stays alive.

Conversely, recombinant NPVs containing genes that encode insecticidal protein rapidly kill the infected insect, thereby reducing the amount of time available to produce occlusion bodies. In a labouratory study conducted at American Cyanamid Company's

Agricultural Research Centre, third-instar H. virescens were dosed with OBs of feral NPV or any of seven different AcMNPV-AaIT clones (differing in promoter-signal sequences within the chimeric toxin-gene cassette).

On average, each of the AcMNPV-AaIT clones killed H. virescens larvae in half the time required by wild AcMNPV, and when cadavers were weighed and assessed for quantity of viral progeny, larvae infected with any of the AcMNPV-AaIT isolates had about five-fold less body mass and contained approximately eight-fold fewer OBs than larvae that were infected with feral NPV.

Numerous published studies have indicated that production of progeny by AcMNPV-AaIT and AcMNPV-LqhIT clones in larvae of H. virescens and T. ni is markedly reduced, in some cases by more than 10-fold, compared with the number of OBs produced by feral AcMNPV in these same host species.

Fuxa *et al.* conducted a series of labouratory and greenhouse tests to compare wild and recombinant forms of AcMNPV for their abilities to propagate and disperse following infection and death of insect hosts. Similar to observations made by other researchers, Fuxa *et al.* found that T. ni survival time and viral reproduction was significantly reduced for insects infected with AcMNPV-AaIT than for those infected with feral AcMNPV.

Further, one week after placement of virus-infected T. ni onto non-treated collard plants in the greenhouse, it was found that larvae infected with wild AcMNPV deposited about 25 times more OBs onto collard foliage than T. ni infected with AcMNPV-AaIT.

*This difference in release of OBs onto crop foliage was attributed to*:

- Fewer OBs being produced in AcMNPV-AaIT infected larvae
- Reduced incidence of cadaver disintegration by AcMNPV-AaIT diseased hosts and a subsequent decrease in liberation of OBs onto collard leaves.

Based upon their findings, Fuxa *et al.* suggested that AcMNPV-AaIT would not be as capable as feral AcMNPV in establishing infections in subsequent hosts, and would therefore be selected against in competition with its feral counterpart in the environment. Indeed, in a caged-plot field trial, Cory *et al.* determined that transmission of virus between successive cohorts of T. ni was significantly lower on cabbage sprayed with AcMNPV-AaIT than in plots of cabbage treated with feral AcMNPV.

A recent labouratory experiment conducted by entomologists and molecular biologists at American Cyanamid Company showed that AcMNPV-AaIT was not competitive with feral AcMNPV in co-infected

H. virescens larvae. In this experiment, individual larvae were initially exposed to a mixture (LD95) of AcMNPV-AaIT and AcMNPV at a 10:1 numeric ratio of OBs, respectively.

Viral OBs produced from this first larval population were extracted from cadavers, pooled and fed to a second cohort of H. virescens. This process of passing viral progeny onto subsequent host-insects continued through six cohort sequences. Restriction enzyme digestion and quantitative DNA blot hybridization were used to determine relative abundance of AaIT-containing viral genomes extracted from each larval passage. Analyses of virus extracted from the initial larval passage showed that 78% of the viral DNA molecules contained the AaIT gene.

However, the viral DNA population collected from the sixth serial passage consisted entirely of feral AcMNPV genotype, i.e. AaIT-containing DNA molecules were not detected in the samples. Such results again suggest that AcMNPV-AaIT is at a competitive disadvantage with feral AcMNPV. Thus, such changes made to the viral genome to make it a more effective insecticide may also be responsible for increasing the likelihood of extinction of the recombinant NPV in the environment.

## REGULATORY AFFAIRS

The US EPA is charged with the regulation of pesticides under both the Federal Insecticide Fungicide and Rodenticide Act (FIFRA) and the Federal Food, Drug and Cosmetic Act as they were amended by the Food Quality Protection Act (FQPA) of 1996. This function is managed by the Agency's Office of Pesticide Programmes (OPP) which is tasked with assuring that the use of a pesticide does not cause unreasonable adverse effects to humans or the environment and that there is a reasonable certainty no harm will result from the aggregate exposure to the pesticide residue (under FQPA).

A pesticide, as defined by FIFRA, is 'any substance ... intended for preventing, destroying, repelling, or mitigating any pest'. This broad definition includes baculoviruses as well as other microbial products being used for control of insect pests. A separate group within OPP was established in 1994 to specifically facilitate the registration of biopesticides such as the baculoviruses and encourage their use. This group, called the Biopesticides and Pollution Prevention Division (BPPD) performs risk/benefit and risk management functions for microbial pesticides among other responsibilities centreed around the use of safer pesticides and IPM programmes. Within BPPD, responsibility is divided among a Microbial Pesticides Branch (which includes plant-pesticides), a Biochemicals Pesticides Branch, and a Pollution Prevention Staff. In order to register a pesticide for use under FIFRA, certain data requirements must be met which allow the Agency to assess the risk to humans and the environment from the use of the product. In general, biopesticides pose fewer risks than conventional pesticides; thus, EPA's data requirements for registration are typically less than those required to register a conventional product.

Since less data need be reviewed by EPA, the time from submission to registration of a biopesticide also is typically less than that needed to register a conventional product. Nonetheless, the Agency is tasked with the conduct of a thorough evaluation of data to ascertain the potential risks. In addition, any pesticide used on food or feed crops in the US must have an established tolerance or, due to the nature of the pesticide, be exempted from the requirement of a tolerance.

A tolerance is the legal maximum residue concentration of a pesticide allowed on food or feed. If a crop is found to have pesticide residues above the legal limit or if pesticidal residues are found on a crop which is not supported by a tolerance, the crop may be considered adulterated and be seized by Federal or State enforcement agencies. Again, since biopesticides are generally considered inherently less harmful than conventional pesticides, tolerance exemptions are typically granted by EPA for microbial products. The data requirements for the registration of microbial pesticide products can be found in the Code of Federal Regulations.

Data required for registration depends upon the proposed use of the microbial pesticide products (i.e. terrestrial food use, forestry use, indoor use, etc.) and includes, but is not limited to, the areas of product analysis, residues, mammalian toxicology, non-target organism toxicology, and environmental expression. Data requirements for an Experimental Use Permit (EUP) under FIFRA 5 are typically a subset of what is required for a full registration under FIFRA 3.

Examples of tolerance exemption regulations for microbial products can be found under 40 CFR 180. Prior to making a development decision and generating data for registration or EUP purposes, the efficacy of a microbial product is typically ascertained under field use conditions. According to 40 CFR 172 Subpart C, field efficacy testing can be accomplished on a small scale (i.e. less than 10 acres for terrestrial crops) without EPA approval for naturally occurring, indigenous microbial products as long as the treated crop is destroyed at the completion of the trial. However, for certain genetically modified microbial products, the Agency requires the submission of a notification of the intent to conduct field trials

Notification is required for small-scale field testing of microbial pesticides 'whose pesticidal properties have been imparted or enhanced by the introduction of genetic material that has been deliberately modified' (i.e. gene-inserted constructs) and for 'non-indigenous microbial pesticides that have not been acted upon by the US Department of Agriculture (i.e. either by issuing or denying a permit or determining that a permit is unnecessary; or a permit is not pending with USDA)'.

Although previously not exempt from the requirement of a notification under 40 CFR 172 Subpart C, a notification is no longer required for field testing genetically modified 'microbial pesticides resulting from deletions or rearrangements within a single genome' (i.e. gene-deleted constructs).

The notification itself must include information and data on the microbial pesticide including but not limited to its taxonomy and natural habitat, means and limits of detection of specific analytical methods, techniques used to genetically modify the microbe, the identity and location of the gene segments inserted or deleted, its physical, chemical and biological features, its host-range and survivability in the environment, along with detailed information on the field trial programme such as rates, specific trial locations and application methods, target pests, site security measures, methods of sanitation of equipment, methods used to monitor movement of the microorganism, and a statement of composition of the experimental formulation.

In addition, the means of detecting potential adverse effects must be submitted along with methods of controlling the microbial pesticide if detected outside the test area. Basically, this information and data are a subset of what would be required for a full registration or EUP for the genetically modified microbial pesticide.

The Agency has established a mechanism in the notification process for protecting data claimed to be confidential business information (CBI). EPA also allows for the petitioning for exemption from the requirement of a notification under 40 CFR 172.52. Once the notification is submitted, the Agency has established a timeframe of 90 days to review the data and respond back to the submitter; however, approval of the notification by EPA must be granted before the field trials can take place.

In addition, the EPA announces receipt of the notification in the Federal Register and thereby opens a public comment period. Upon review of the notification and public comments the Agency can:

- Approve the proposed test;

- Approve the proposed test with certain modifications to the programme;
- Request additional information from the submitter;
- Require an EUP be submitted under FIFRA 5;
- Disapprove the proposed test and any EUP application. Following Federal approval of the field testing of the microbial pesticide, individual states (i.e. California) may have their own regulatory approval processes in place and the appropriate State regulatory authorities should be consulted prior to conducting field trials.

As part of its programme to develop recombinant baculovirus biopesticide, American Cyanamid Company first submitted a notification to conduct small-scale field testing of its egt gene-deleted Autographa californica MNPV construct in 1993. Although this gene-deleted construct does not require a notification under the current regulations, at the time of submission, notifications were indeed required for gene modifications of this type.

Based on information submitted with the notification in conjunction with literature from the public domain, the Agency concluded that 'no unreasonable adverse effects on human health and environmental effects from the conduct of this small-scale field test are foreseen'.

EPA approved the conduct of the field trial programme with a soil-monitoring requirement and feeding of the sampled soil to the highly permissive H. virescens larvae. The Agency required that any larvae showing signs of viral infection after being exposed to the treated soil must be examined for occlusion bodies (OBs) and any OBs found must be analysed by polymerase chain reaction assay (PCR) to determine if recombinant DNA is involved.

The following year, American Cyanamid Company submitted a notification of intent to conduct small-scale field testing of its AaIT gene-inserted Autographa californica MNPV (AcMNPV-AaIT) construct. Upon review of the submitted data and in response to public comment, the Agency requested more information concerning the environmental expression (i.e. persistence in the environment) and host-range of the gene-inserted baculovirus.

After several rounds of meetings and correspondence with EPA personnel aimed at addressing their concerns, the initial notification for field testing was ultimately approved by the Agency, albeit with substantially increased mitigation (i.e. limited acreage, site disinfection procedures, soil barriers, 'wild-type' baculovirus oversprays), increased monitoring (i.e. soil analyses for OBs, non-target insect surveys), and decreased acreage requirements in comparison to the gene-deleted construct. The mitigation and monitoring requirements were aimed at addressing the questions on persistence and host-range and more fully characterizing any potential risk. As the initial field trials were underway, scientists at American Cyanamid Company continued to conduct labouratory studies focused on the host-range determination by looking at various species of lepidopteran and non-lepidopteran insects.

Additionally, a 'competition' study was conducted which proved that the ecologically disadvantaged modified baculovirus (i.e. recombinant produces less OB progeny than the wild-type virus) would not be able to overcome the naturally occurring virus in the environment. These studies, in conjunction with the results from two years of field monitoring work, provided the Agency with sufficient data to ultimately reduce the mitigation and monitoring requirements in follow-on notifications.

By 1997, the Agency had monitoring data collected from the 1995 and 1996 trials which, along with numerous labouratory bioassays, showed that AcMNPV-AaIT had no effect on non-target arthropods (at least 26 families of Insecta and Arachnida representing 45 species were encountered in monitoring surveys with no ill-effect observed). In addition, soil monitoring data from the earlier trials showed that the recombinant 'does not vary in its environmental fate (i.e. persistence) profile in comparison to the naturally occurring Autographa californica MNPV'.

The Agency subsequently approved a small-scale field programme for the 1997 season and removed the requirements for soil/non-target insect monitoring and site disinfection. In fact, EPA no longer requires notification for field testing of AcMNPV-AaIT as long as certain minimal mitigation criteria are followed. Also in 1997, American Cyanamid Company submitted a notification for small-scale, crop-destruct field testing of an AaIT gene-inserted construct based on a H. zea NPV baculovirus.

This notification built on the experience gained from the modified AcMNPV field work and included labouratory-derived results which once again showed that neither the host-range nor the infectivity of the recombinant H. zea NPV changed in comparison with the wild-type baculovirus. The Agency approved the small-scale field trial programme and stated that no further notifications were necessary for similar constructs of AaIT gene-inserted H. zea NPV 'if there is no indication of increased host-range' as long as the tests are not conducted in areas adjacent to potentially susceptible endangered insects and that wild-type H. zea NPV baculovirus is introduced to the site as part of a standard treatment or overspray. According to EPA, this overspray of the trial site is a valid way to provide a source of wild-type baculovirus to compete with the genetically engineered baculovirus.

In a notification submitted in 1998, American Cyanamid Company requested approval of small-scale field testing of a tox34 gene-inserted AcMNPV (AcMNPV-tox34). Expression of the tox34 gene in recombinant AcMNPV significantly improved the efficacy of the baculovirus in the same fashion as expression of the AaIT gene in the earlier AcMNPV constructs.

This submission once again provided EPA with information on the host-range and infectivity of the construct. As this tox34 gene-insert construct introduced an insect-specific toxin not previously reviewed by EPA, labouratory data were submitted to ascertain the mammalian toxicity characteristics of the toxin expressed in infected target insect larvae. No adverse toxicological effects were seen. The notification was approved (and further notifications of the tox34 construct waived) with the same wild-type baculovirus overspray restriction and infectivity testing as the AaIT gene-insert H. zea NPV construct.

Overall, EPA concerns with small-scale field testing programmes of genetically modified baculoviruses focused on potential risks of an increase in host-range. Because of the novel technology and after a thorough review of the submitted host-range and environmental safety data, EPA approved the initial small-scale field programms of the original recombinant baculoviruses with extensive risk mitigation criteria. As the database on host-range, infectivity and environmental expression increased, the Agency was able to rely on past studies and extensive monitoring as a foundation to support newer construct introductions.

This increase in familiarity with the science ultimately led to the approval of small-scale field testing of several different recombinant constructs with lessened mitigation requirements and laid the groundwork for further development work. Baculovirus safety in conjunction with recombinant technology should be perceived as having two levels: level 1, species specificity and level 2, target specificity.

The safety of feral baculoviruses has been clearly established through several decades of high-quality research. The first level of safety is dictated by the fact that infection can occur only in the limited number of species susceptible to any baculovirus. In the recombinant form of the virus, a second level of safety is added, in that the inserted gene encodes a toxin which binds only to insect targets.c Research on the recombinant viruses is somewhat limited since their availability is a recent development. The constructs currently available meet the two-level safety thesis. The limited database does show that these constructs do not infect species outside their established host-range and, in fact the addition of the toxin gene does not even alter the LD50.

The major benefit is a faster kill and better crop protection (desirable changes). Very sophisticated cellular level experiments have shown that there is no replication in non-target hosts (vertebrate and invertebrate). Finally, the recombinant NPV is clearly at a competitive disadvantage to the feral NPV due to its limited ability to propagate. The growing database has been well received by rational environmental groups, academics and growers. USA regulatory agencies are becoming convinced about the safety of these recombinants and see them taking their place in the grower's arsenal of insect control tools.

## Chapter 8

# Foods Containing Genetically Modified Materials

### DNA TRANSFER

The first food products containing genetically modified material are now available on supermarket shelves throughout the world. Much public debate has arisen concerning the safety of such products and indeed the need for genetically modified foodstuffs in the well-stocked larders of the Western World.

However, because genetic engineering offers such technical advantages to the food industry in the mass production of cheap processed food of predictable consistency and quality, there will be increased commercial pressure to broaden the range of genetically modified foodstuffs available in the marketplace.

Similarly, as our confidence in the safety of this new technology grows and our ability to apply genetic engineering to products which offer real benefits to the consumer (e.g. safer, more wholesome food) and in the development of 'functional' foods which may well play an important role in human health and disease prevention, consumer acceptance and demand for such products may grow.

However, consumer confidence will rely on rigorous assessment of the various potential risks involved and appropriate safety testing. Microorganisms, particularly lactic acid bacteria, have been used in the production of fermented foods for millennia. Recent advances in genetic engineering allow, for the first time, accurate identification of microorganisms traditionally used in food fermentation and the design of novel strains with improved characteristics.

Age-old production problems such as instability of industrially important traits and failure of starter cultures due to bacteriophage attack may now be tackled at the molecular level. It has long been proposed that certain lactic acid bacteria or 'probiotics' contribute greatly to intestinal health and well-being. Similarly, genetically modified lactic acid bacteria have also been proposed for use as oral vaccines. Recent advances in molecular microbial ecology allow the scientific basis of such claims to be determined and genetic engineering will enable the design of probiotic bacteria with specific health-promoting properties. No food products containing live genetically modified microorganisms (GMMs) are available in the marketplace at present.

However, commercial pressure on the food industry and the consumer benefits promised by probiotic strains designed with scientifically proven health-promoting capabilities will encourage their use in fermented foods in the near future. Clearly the biosafety of such products must be rigorously investigated before they become commercially available, particularly in view of the public back-lash towards genetically modified plant material used in the food products in Europe.

Of particular concern when considering the release of live genetically modified microorganisms in food is the possibility of recombinant DNA transfer from genetically modified organisms (GMOs) to members of the human gut microflora. Transfer of DNA between bacteria occurs naturally in the environment and offers prokaryotes a unique means of evolution and adaptation in response to changing environmental conditions.

*Interest in DNA transfer between bacteria in the mammalian gastrointestinal tract stems from three main areas*:

- The possibility of DNA transfers from GMOs which may be ingested in food to members of the human gut microflora.
- The spread of antibiotic resistance amongst bacteria as a result of gene transfer.
- The emergence of novel human pathogens as a result of transfer of virulence factors or antibiotic resistance determinants between bacteria.

The focus of this review is to discuss the ability of bacteria to undergo DNA transfer in the human gastrointestinal tract with particular reference to the risks posed by genetically modified microorganisms, which may be used in the production of fermented foods such as bread, beer, cheese and yoghurt. We will look at the existing procedures available to monitor DNA transfer in the human gut microflora and investigate some of the factors governing frequencies of such transfer events.

These studies will provide information relevant to our understanding of the possibility of the transfer of marker genes used in the construction of genetically modified crops to the bacteria present in the mammalian gastrointestinal tract. First let us look at the mechanisms of DNA transfer available to bacteria in natural environments.

## TRANSFORMATION

Transformation is the process by which a naked piece of DNA from the environment binds to the surface of a competent bacterial cell and is taken up by the bacterium. DNA may then be incorporated into the host genome, depending on the recombinational abilities of the host and the 'foreign' DNA. The ability to translocate DNA across the cell boundary is called competence.

Competence is a specific physiological state of a bacterial cell (genetically encoded in some cases, e.g. Bacillus subtilis), which occurs transiently and is restricted to certain stages of the growth cycle. Natural competence has been observed in bacteria from a variety of genera, including Haemophilus, Neisseria, Streptococcus, and Bacillus, Acinetobacter and Pseudomonas spp. and Helicobacter pylori.

Recalcitrant species, such as Escherichia coli, may be rendered competent using a variety of chemical, enzymatic and physical procedures, e.g. $CaCl_2$ treatment and electroporation. Such physiochemical conditions may sometimes be prevalent in the local environment of recalcitrant bacteria. For example, $Ca^{2+}$ concentrations in drinking water sometimes approach the levels used in vitro to induce a state of competence in E. coli.

Competence is not the only bacterial-encoded parameter shown to play a role in natural transformation. In some cases, requirements for specific lengths of DNA, DNA states (double or single stranded) and the presence of specific DNA sequences have been observed. Competent Bacillus and Streptococcus spp. may take up any piece of DNA but only homologous DNA will be maintained in the bacterial genome.

Haemophilus spp., on the other hand, requires the presence of specific 11-bp sequences before DNA uptake occurs. These 11-bp recognition sequences occur at a number of locations on the Haemophilus genome. It has also been reported that a diffusible factor may play a role in the induction of competence in Streptococcus spp. Here, induction of competence was found to be dependent on cell density and the pH of the surrounding environment.

### Factors Affecting Transformation in Natural Environments

The presence of naked DNA has been demonstrated in a number of natural environments. Naked DNA may arise in a given habitat via a number of routes. Cell lysis as a result of cell death or the activities of bacteriophage will release bacterial DNA and DNA may be released from actively growing bacteria during certain stages of the growth cycle.

The uptake of DNA from the environment will not only depend on a state of competence in the bacterium but also on the persistence of naked DNA in a given environment. In this respect, different microhabitats will vary greatly in their abilities to either protect naked DNA by the presence of favourable salt concentrations or absorption onto solid supports (e.g. the surface of soil particles or food particles in the gut) or to degrade DNA, for example, by active nucleases present in the microenvironment.

Despite the longevity of DNA in the soil and the presence of bacteria with known competence for transformation, gene transfer from plant to soil bacteria appears to be an extremely rare event: for example, transfer of an ampicillin resistance gene from a transgenic potato line to the plant pathogenic bacterium Erwinia chrysanthemi was at a calculated frequency of 2 × 10-17 and from transgenic plants to the soil bacterium Acinetobacter calcoacetinus at a frequency lower than 10-13. However, the uptake and integration of transgenic plant DNA via natural transformation has been shown for Acinetobacter sp., strain BD413 by in vitro marker rescue using DNA from various transgenic plants containing the bacterial kanamycin-resistance gene nptII. Naked DNA enters the human gastrointestinal tract from a number of sources. These include ingested food and foreign microorganisms as well as members of the human gut microflora.

Very little is known about the ability of naked DNA to persist in the gut and evade the activities of mammalian and bacterial nucleases. Factors which may affect the persistence of naked DNA, and thus the incidence of transformation in the human gastrointestinal tract, include pH, salt concentrations, cell densities, local nuclease activities and protection afforded by absorption onto surfaces such as food particles or mucosal surfaces.

Such factors would act at the level of the microhabitat and as such would vary greatly even within specific regions of the gut. It has been shown that the CRY1A protein and DNA from genetically modified maize are digested by simulated gastric juices in vitro. However using an in vitro model of the intestinal tract, van der Vossen *et al.* showed that 6% of transgenic tomato DNA survived the stomach and small intestine and concluded that the presence of raw mashed tomato helped to preserve the DNA. Free chromosomal DNA of Bacillus subtilis persists for weeks in milk and dairy produce. Such bacteria develop natural competence and can be transformed with free chromosomal or plasmid DNA in such produce. Schubbert *et al.* showed that the wall of the gastrointestinal tract is exposed to a variety of DNA fragments and remains exposed to DNA fragments of dietary origin for hours after ingestion of the food.

The authors found that upon feeding mice 50 μg M13mp18 DNA, approximately 95% of ingested DNA was lost during passage through the stomach. However, phage DNA could be detected by PCR and fluorescent in situ hybridization in peripheral leukocytes, spleen and liver cells as well as the contents of mouse small intestine, caecum, large intestine and faeces.

Ingested phage DNA was detected for up to 18 hours after ingestion in caecal contents, for up to 8 hours in DNA from the peripheral blood cells and for up to 24 hours but not 48 hours in DNA from spleen and liver. Mercer *et al.* monitored the survival of recombinant plasmid DNA (pVACMC1) in fresh human saliva. The fraction of naked DNA remaining amplifiable in saliva ranged from between 40 and 65% after 10 minutes and 6-25% after 60 minutes. Amplifiable plasmid DNA was still present after 24 hours incubation in fresh saliva.

The authors also found that plasmid DNA, which had been exposed to degradation by saliva, was capable of transforming naturally competent Streptococcus gordonii DL1 in filtered saliva. Transformation activity decreased rapidly with the extent of plasmid DNA degradation.

Such studies suggest that DNA released from food or bacteria ingested with food may undergo transformation in not only the oral cavity but other regions of the human gastrointestinal tract. Clearly, much more research is needed on the ability of naked DNA to persist in the lower regions of the human gastrointestinal tract and the extent to which transformation contributes to DNA transfer in the human gut microflora.

## TRANSDUCTION

Transduction is the mechanism by which DNA may be transferred between bacteria by bacteriophage. In essence, the bacteriophage acts as microbial couriers, picking up DNA from one bacterial chromosome and delivering the heterologous DNA to another bacterial chromosome.

Where degradation of naked DNA is one of the chief factors limiting DNA transfer by bacterial transformation in natural environments, bacteriophage protects DNA in natural environments via their proteinaceous capsid. Bacteriophages, on the whole, contribute to gene transfer between bacteria by two main mechanisms, namely specialized and generalized transduction.

### Specialized Transduction

Specialized transduction involves the incorporation of a lysogenic phage into the bacterial chromosome. Lysogeny is favoured by conditions of environmental stress such as the limitation of nutrients and probably aids the survival of the phage and host bacterium. Upon excision from the chromosome, elements of host DNA adjacent to the prophage DNA may become excised along with the phage DNA. The host DNA may then become incorporated into the bacteriophage genome and packaged along with the phage DNA.

Thus, when infection of a recipient bacterium occurs the heterologous DNA may become integrated into the recipient's chromosome during lysogeny.

### Generalized Transduction

Generalized transduction on the other hand, occurs when host DNA is packaged into phage particles instead of the phage genome. This mechanism of transduction has been observed in the lytic phage, P1 and P2 of the enterobacteraceae, for example. Upon transduction to a novel recipient, the transferred DNA may either be incorporated into the host chromosome via recombination or where it possesses the means of self-replication, it may replicate autosomally.

### Factors Limiting DNA Transfer by Transduction

Transduction is greatly dependent on the host range of the transducing bacteriophage, which is generally narrow, since bacteriophage infection is dependent on the phage recognizing specific receptor sites on the bacterial cell surface. Some bacteriophage that are able to mediate transduction between different species of bacteria have been described, e.g. P1 and Mu.

The frequency of transduction in nature may be much higher than previously recognized, since the number of bacteriophage particles in many environments appears to be much higher than first thought. Whether the transferred DNA is maintained in the new host is greatly dependent on the ability of the bacteriophage to insert itself and the heterologous DNA into the host genome and evade the bacterial restriction modification system. Bacterial restriction enzymes may recognize specific sequences on heterologous or phage DNA. It has been proposed that restriction modification systems may reduce infection of unmodified phage DNA by 2-3 orders of magnitude. The amount of DNA that may be transferred by transduction is also limited, being about equivalent in size to the phage genome itself.

## Transduction in the Human Gastrointestinal tract

Despite the fact that the bacteriophage are ubiquitous members of natural microbial ecosystems, little is known about their distribution or activity in the human gut microflora. To date no reports of transduction in the human gastrointestinal tract have been presented. However, where sufficient effort has been made to isolate bacteriophage from natural environments and examine their activity in situ, they have been found to play a significant role in microbial ecology.

The fact that about 90% of all bacteriophage isolated from natural environments are temperate suggests that transduction may be more important in microbial genetic plasticity than previously appreciated. Thus specifically designed studies involving in vitro or in vivo models of the human gastrointestinal microflora may well provide evidence of transduction in this ecosystem and elucidate some of the ecological factors governing transduction.

Bacteriophage specific for major groups of bacteria present in the gut microflora, e.g. Bacteroides spp., bifido-bacteria, lactobacilli, methanogens, clostridia, enterobacteriaceae, streptococci and staphylococci have been identified. However, transduction has not been observed in many of these bacterial groups e.g. Bacteroides spp. or Clostridium spp.

Bacteriophage of the lactic acid bacteria play a major role in the dairy industry both as destructive agents, causing the failure of starter cultures in cheese and yoghurt production, and as valuable genetic tools in the development of genetically altered industrial strains. The variety of bacteriophage associated with the lactobacilli suggests that transduction may be a significant means of gene transfer within these genera.

Tohyama *et al.* demonstrated that the Lactobacillus salivarius temperate phage PLS-1 mediates generalized transduction of auxotrophic markers (lysine, proline and serine) and lactose metabolism at frequencies of 10-7 to 10-8 transductants per CFU in vitro. Later, Raya *et al.* showed that phage ADH replicates in a lytic cycle, establishes lysogeny, confers superinfection immunity on the host and mediates plasmid DNA transduction in L. acidophilus ADH. It was also observed that plaque formation on cell lawns of L. acidophilus NCK102 was pH dependent, with the optimal pH for plaque formation being pH 5.5. Transfer of antibiotic resistance determinants have been reported between strains of Desulfovibrio desulfricans via phage Dd1 mediated transduction at frequencies of 10-5 to 10-6 transductants per recipient.

Transduction has also been observed in methanogens, although not strains found in the human gut microflora. Despite demonstrations that transduction does occur under labouratory conditions, little is known about the significance of transduction-mediated gene transfer in the natural and animal-associated environments.

Important questions remain unanswered regarding the prevalence and survival of bacteriophage in different environments, the limitations imposed on transduction by the host specificity of the transducing bacteriophage and the frequencies at which transduction occurs during phage replication.

## BACTERIAL CONJUGATION

Plasmid DNA, extrachromosomal, self-replicating genetic elements, may mediate DNA transfer between bacteria via conjugation. Conjugation involves cell-to-cell contact between a donor (plasmid-bearing) and recipient (plasmid-free) cell. Not all plasmids are conjugative but many conjugative plasmids carry environmentally important traits such as antibiotic resistance determinants, virulence factors, novel degradative pathways.

Different bacteria or groups of bacteria display different mechanisms of conjugation, a comprehensive discussion of which is beyond the scope of this review. However, a short, generalized description of some of the conjugative mechanisms employed by different groups of bacteria may serve to highlight both the mechanistic diversity of bacterial conjugation and that DNA transfer via conjugation-like mechanisms has been observed in bacteria from a wide phylogenetic background.

### Conjugation in Gram-negative Bacteria

In Gram-negative bacteria, cell-to-cell contact between a donor and recipient cell is accomplished via the formation of a pilus (which can be short and rigid or long and flexible) by the donor cell. This pilus, a proteinaceous tube, forms a cytoplasmic bridge between donor and recipient, along which a single-stranded copy of the conjugative plasmid is transferred to the recipient cell.

DNA replication then leads to generation of the complementary strand of plasmid DNA in both recipient and donor. Certain plasmids originally found in Gram-negative species display an extremely broad host range. Such plasmids (e.g. the R plasmids) may carry a number of antibiotic-resistance determinants and have been shown to transfer to a variety of bacterial species in different natural environments. Some conjugative plasmids may mediate transfer of chromosomal DNA between donor and recipient cells upon integration into the chromosome via homologous recombination and low-fidelity excision from the chromosome, leading to incorporation of segments of bacterial DNA into the circularized plasmid before transfer.

Transfer of Gram-negative plasmids has been observed in many natural environments, including soil, rhizosphere, salt and fresh water and effluent.

## Conjugation in Gram-positive Bacteria

Conjugation of plasmids in Gram-positive bacteria differs considerably from those in Gram-negative bacteria. Two main mechanisms have been described. Certain plasmids found in Enterococcus spp. and Lactococcus spp. have been shown to initiate conjugation in response to a small peptide signal released extracellularly by plasmid-free recipient cells.

These extracellular peptides act as sex pheromones, and lead to clumping of donor and recipient cell to form a conjugative aggregate. This increases cell-to-cell contact between donor and recipient cells and results in the transfer of pheromone-induced plasmids at high frequencies.

Plasmid pAD1 of Enterococcus faecalis is the best characterized example and is transferred at high frequencies in conjugative aggregates formed in response to pheromones released by plasmid-free recipient cells. The second type of conjugation elucidated in Gram-positive bacteria, especially Streptococcus spp. and Staphylococcus spp., depends on the direct contact between donor and recipient cell imposed by growth on a solid surface.

Plasmids employing this form of DNA transfer are not induced to conjugate by sex pheromones and the mechanism of their transfer remains unclear. Such plasmids often display a remarkable broad host range, i.e. they can be transferred to a wide spectrum of both Gram-positive and Gram-negative bacteria, including Streptococcus spp., Staphylococcus spp., Clostridium spp., Lactobacillus spp., Listeria spp., Pediococcus spp. and Bacteroides spp. Many of these plasmids encode one or more antibiotic-resistance determinants and are thought to have played a role in the dissemination of antibiotic-resistance determinants among bacteria of clinical importance. The best characterized examples include pAMâ1 and pIP501.

## Conjugation in Archaea

Little is known about the molecular biology of Archaea because of the difficulties in cultivating such organisms routinely under labouratory conditions and the major differences that exist between Archaea and Bacteria. However, a conjugation-like mechanism of DNA transfer has been observed in the halophilic Archaea Halobacterium volcanii.

Here, a 'cytoplasmic' bridge forms between parental cells along which chromosomal DNA has been shown to transfer. The mechanism of DNA transfer displays characteristics of both bacterial conjugation and eukaryotic cell fusion. However, until we gain a greater understanding of the molecular biology and indeed microbiology of the Archaea, few conclusions can be drawn about the frequencies or mechanisms of DNA transfer in Archaea.

## Conjugative Transposon

Transposons, genetic elements borne by microbial genomes and possessing the ability to 'transpose' between different locations on the same genome, have long been known to play a role in microbial genetic plasticity. Some such elements not only possess the machinery required for transposition but are also capable of mediating their own transfer between the genomes of different bacteria via conjugation. Such conjugative transposons have been found in a wide array of both Gram-positive and Gram-negative bacteria.

They are thought to play a significant role in the dissemination of antibiotic-resistance determinants among a wide range of bacteria and the emergence of multiple antibiotic-resistant pathogenic strains. Conjugative transposons generally excise from the bacterial chromosome to form a covalently closed circular double-stranded DNA transposition intermediate.

The transposition intermediate can either reinsert itself into the bacterial chromosome or onto a plasmid in the same cell, or it can mediate its own conjugation to a second bacterial cell and integrate into the recipient chromosome. Conjugative transposons have a wide host range and have been shown to transfer between bacteria at high frequencies typically 10-5 to 10-4 per recipient.

## Mobilization

Some conjugative plasmids may 'mobilize' plasmids co-resident in a bacterial cell. Mobilization results in the transfer of a non-self transmissible plasmid from one bacterial cell to another and is mediated by the conjugative plasmid or conjugative transposon. Mobilization may occur as a result of the non-conjugative plasmid 'hijacking' the conjugative machinery of the conjugative plasmid or the two plasmids may form a co-integrate plasmid. Plasmids with oriT sequences may be mobilized upon formation of efficient cell-to-cell contact between donor and recipient bacteria mediated by the trans acting products of tra and mob genes borne on self-conjugative plasmids co-resident in the same cell. Plasmids devoid of oriT sequences may be mobilized by formation of a co-integrate with a self-conjugative plasmid co-resident in the same cell. Such co-integrate formation is characteristic of bacterial transposable elements. The co-integrate once transferred to a recipient strain via conjugation, may then resolve into its constituent plasmids. Mobilization of genetically modified plasmids has been demonstrated from introduced genetically modified donor strains to bacteria present in different environments. Such observations have implications for the release of genetically modified microorganisms into natural environments, even when steps have been taken to limit the transfer potential of recombinant DNA.

Many broad host range plasmids are also mobilizable, which aids their dissemination within microbial consortia as well as the dissemination of antibiotic resistance often borne on such plasmids. Some conjugative plasmids can also mobilize chromosomal DNA between different bacteria. Conjugative transposons have also been shown to mediate mobilization of non-conjugative plasmids or chromosomal genetic elements between bacteria.

## RETROTRANSFER

Transfer of plasmid DNA between bacteria is usually unidirectional, i.e. from donor to recipient. However, Mergeay *et al.* demonstrated the transfer of genetic markers from a recipient strain to the donor strain during conjugation experiments. This form of plasmid transfer has been called retrotransfer. The mechanism involved has not been fully elucidated, but is thought to involve multiple donor and recipient cells. Retrotransfer has been Demonstrated in a number of bacterial species.

### Factors governing DNA Transfer

Plasmid transfer by conjugation is determined by characteristics of the plasmid itself, e.g. tra+, mob+ and oriT+ genotypes, and the donor and recipient bacteria involved. Environmental parameters which affect pilus formation or the rate of plasmid curing, such as SDS, organic solvents, proteases, high temperature and acridine orange, will affect the frequency of conjugation. The metabolic character of the host bacterium, e.g. growth rate or stress, may also affect the frequency of conjugation. The frequency of conjugation may be enhanced by the presence of surfaces and bacterial aggregation (e.g. mediated by bile salts or sex pheromones). The presence of solid surfaces, e.g. soil particles, food particles or walls of mucosa, in an environment may also affect the frequency of plasmid transfer depending on the mode of conjugation employed, by increasing cell-to-cell contact between donor and recipient cells. Substrate availability may also affect conjugation with increased frequencies of conjugation sometimes observed with a plentiful supply of bacterial substrates.

Physiological conditions extant in the environment, such as pH, temperature, substrate availability and bacterial cell density, may all play a role in governing the frequency of conjugation in a plasmid- and bacterium-specific manner. Plasmid maintenance by bacteria exacts a metabolic cost on the cell, i.e. the energy needed for the plasmid to replicate and express plasmid encoded proteins.

This 'metabolic burden' may be counteracted by adaptational advantages endowed by plasmid maintenance. Plasmids often encode environmentally important traits such as resistance to antibiotics and bacteriophage, unusual metabolic degradative pathways, virulence factors and genes encoding anti-bacterial compounds. Such traits may enable a plasmid-bearing cell to out-compete a plasmid-free cell in response to altering environmental conditions, e.g. presence of antibiotics, unusual substrates or bacteriophage.

Similarly, environmental factors affecting the competitiveness or growth rate of a bacterial cell will affect the ability of the cell to both undergo conjugation and maintain newly acquired plasmids. Properties of the plasmids themselves may limit the extent to which conjugation occurs in the environment. For example plasmid incompatibility, host range, surface exclusion, evasion of host restriction modification systems and segregational stability may all play a role in the extent to which a particular plasmid may spread and persist in natural microbial communities.

Thus it appears that the frequency of plasmid transfer in a given environment will be a function of the interaction between environmental factors and the bacterial populations involved, characteristics of the plasmid itself and the mode of conjugation employed by the plasmid.

## TRANSFER OF DNA FROM GMO

There is potential for the presence of live GMOs in foodstuffs such as yoghurt and fermented sausage produced with starter cultures containing genetically modified organisms or perishable foods treated with protective cultures of GM organisms. The environment within living food may permit gene transfer by conjugation or transduction and this has been seen at high frequency in complex food matrices.

Free recombinant DNA may also be present in food products resulting from the killing of GMMs during food processing. The human gut microflora is extremely complex, being made up of more than 500 different species of bacteria and climax microbial populations reaching up to 1011 CFU/g luminal contents in the colon. The high density and species diversity of bacteria present in the colon provides ideal conditions for the transfer of DNA between bacteria.

However, the diversity of the human gut microflora also presents significant problems in monitoring the DNA transfer events between bacteria and, in particular, monitoring the fate of recombinant DNA, which may be ingested in food. It has been determined that only a fraction of the bacteria present in the human gut microflora may be cultured under labouratory conditions and characterized.

Many species found in the human gut microflora defy culture by standard microbiological techniques, and selective agars and growth conditions exist for only a small proportion of the predominant species present. Much progress as been made in recent years in the development of culture-independent methods for identifying bacteria present in natural environments.

Using the phylogenetic information present in 16S rRNA and fluorescent in situ hybridization techniques, it is now possible to identify bacteria in situ in natural environmental samples including human faeces. However, such techniques tell us little about the genetic capabilities of bacteria and by themselves do not allow us to monitor the distribution or indeed dissemination of specific genetic determinants amongst natural assemblages of bacteria.

Thus, studies on the transfer of DNA between members of the human gut microflora and from GMOs to members of the human gut microflora have employed specific marker systems borne on the mobile genetic elements or on recombinant DNA. Traditional marker systems used in pure culture microbiology such as the lac operon or X-gal are of little use when employed to monitor DNA transfer events in complex assemblages of gut microorganisms, because of the widespread occurrence of such genes amongst the mammalian gut microflora.

Antibiotic-resistance determinants, when shown to be absent from background bacteria, have been widely employed in labouratory-based studies. However, such marker systems cannot be used in studies with human volunteers because of the emergence of antibiotic-resistance determinants in pathogenic strains. Novel marker systems such as green fluorescent protein are now being employed in labouratory studies to monitor the persistence of GMOs in the human gut microflora and DNA transfer events between bacteria. Another hurdle to monitoring the dissemination of recombinant DNA in complex microbial consortia and investigating the factors governing DNA transfer in natural environments is the non-expression of marker genes in novel bacterial hosts.

Marker genes transferred between bacteria may not always be expressed in the recipient cell. Thus not all DNA transfer events will be detected by existing marker systems. One way around this problem is to use a specific plasmid transfer system comprising a donor and recipient bacterium in which the marker gene is known to be expressed, and then examining the factors governing DNA transfer between these two strains in models of the ecosystem under investigation.

Recent advances in molecular biology offer some hope that more reliable means of monitoring the dissemination of genetic determinants amongst diverse bacteria may be developed. Fluorescent in situ hybridization is now being employed to enumerate bacteria in natural assemblages by hybridizing 16S rRNA targets to fluorescently labelled DNA probes and visualization with fluorescent microscopy. Similarly, it is now possible to visualize PCR products within bacterial cells fixed on microscope slides.

By combining these techniques it may soon be possible to visualize concomitantly, fluorescently labelled PCR products of single-copy genes and labelled 16S rRNA sequences within a single cell in environmental samples. To enable us to ask specific questions about the human gut microflora, in vitro and in vivo models of the human gut have been developed.

In vitro models generally employ anaerobic chemostats of varying degrees of complexity or short-term batch cultures of human faeces or luminal samples extracted from animals. They allow generation of data on the effect of specific biotic and abiotic environmental parameters on the gut microflora, which may otherwise be unavailable due to inaccessibility of the target environment in the gastrointestinal tract or for ethical reasons.

Freter *et al.* studied the transfer of plasmid DNA between E. coli strains using a single-stage continuous flow culture model of the mouse gut microflora. This in vitro model of the mouse caecal microflora simulated microbial interactions observed in the mouse gut. Data generated from the in vitro model also allowed validation of mathematical models of plasmid transfer between E. coli strains in the presence of the complex mouse gut microflora. Rang *et al.* employed short-term batch cultures of luminal extracts from the mouse gastrointestinal tract to monitor transfer of plasmid RP1 between E. coli strains. More complex in vitro models of the human gastrointestinal tract developed recently have yet to be employed to monitoring DNA transfer events in the human gut microflora.

In vitro models allow us to examine interactions between specific members of the gut microflora or between the whole human faecal microflora and introduced GMOs and non-GMOs under the physiological conditions of the human gut.

They provide a vital first stage in the biosafety assessment of GMOs intended for use in human food. However, such models do not take into consideration the biological complexity of the mammalian gut ecosystem. In vitro models may not simulate the biological surfaces or the input of endogenous substrates and secretions of the human immune system, which may play a vital role both in the maintenance and establishment of the gut microflora, and the frequency of DNA transfer events.

Gnotobiotic and germ-free animals have been employed as in vivo models of the human gastrointestinal ecosystem for some years. Gnotobiotic technology provides a useful tool in studying the interactions between specific members of the human gut microflora, e.g. the role played by specific members of the microflora in colonization, resistance to pathogenic bacteria, bacterial pathogenesis, and metabolism of dietary constituents or xenobiotic compounds.

Transfer of plasmid DNA between specific members of the human gut microflora has also been studied using di- or poly-associated animals. Conventional animals may not provide information of direct relevance to the human gut ecosystem because of the major differences in the composition of the gut microflora between animals and man. To circumvent these problems, germ-free animals have been associated with groups of bacteria from the human gut microflora or the whole human gut microflora.

Such human flora-associated (HFA) animals provide an essential tool in studies of human gut microbiology in that they reflect the complexity of the human gut ecosystem in terms of microbial species and numbers, and they provide some of the biotic parameters provided by the mammalian mucosa, immune system and digestive functions.

Such in vivo models provide an extremely useful tool in the generation of data on the safety of GMOs in food until a greater degree of confidence in the safety of such products is attained and human volunteer studies can ethically be considered.

## Factors Affecting Transfer of Plasmid DNA from Bacteria in vivo

Very little is known about the extent to which transformation and transduction contribute to DNA transfer between bacteria in the human gastrointestinal microflora. Only one study on bacterial transformation in fresh human saliva samples has been presented in the literature. No reports of transformation or transduction in lower regions of the human gastrointestinal tract were found after exhaustive search of the literature.

This is probably due to the fact that conjugation has been viewed as the dominant contributor to DNA transfer between bacteria in natural environments, because of the ubiquitous occurrence of conjugative elements amongst diverse bacterial species. Thus most studies looking at DNA transfer in the gut have concentrated on conjugation.

Earlier, it was proposed that the frequency of plasmid transfer in a given environment will be a function of the interaction between environmental factors and the bacterial populations involved, characteristics of the plasmid itself and the mode of conjugation employed by the plasmid. Some of these parameters have been studied in vivo in animal models.

## Effect of Antibiotics on Plasmid Transfer

A number of early investigators observed the transfer of plasmid DNA between bacteria in the gastrointestinal tract of human volunteers. However, many of these studies were conducted with antibiotic-resistance determinants in E. coli strains and in the presence of antibiotic selective pressure. In the light of the recent emergence of pathogenic bacteria with multiple antibiotic resistance determinants it is no longer considered ethically justifiable to carry out such investigations in human volunteers.

Ingestion of antibiotics leads to perturbation of the natural homeostasis of the human gut microflora and presents a selective pressure for the evolution of antibiotic resistance strains through mutation and transfer of resistance determinants. The human gut microflora is commonly exposed to antibiotics, both through clinical practice and use of antibiotics in animal husbandry. Similarly, antibiotic-resistant bacteria have been isolated from a wide range of human foodstuffs.

Duval-Iflah *et al.* observed the transfer of an R plasmid from Serratia liquifaciens to an E. coli recipient strain, naturally present in the human microflora of the HFA mice. The donor S. liquifaciens strain became established in the HFA mice at between 106 and 107 CFU/g faeces. Transfer was observed 12 days after inoculation with the donor strain and their numbers ranged from between 103 and 105 CFU/g faeces. The number of transconjugants greatly increased after addition of selective antibiotic in drinking water. Morelli *et al.* showed that a subtherapeutic dose of selective antibiotic (erythromycin at 10 μg/ml drinking water) increased the level of pAMα1 transfer from L. reuteri to Ent. faecalis from 1 × 10-7 to 1 × 10-4 transconjugants per donor in anexic mice (i.e. gnotobiotic mice associated with donor and recipient strains).

Although selective antibiotic pressure previously has been employed to demonstrate transfer of genetic elements between bacteria in animal models and in humans, the observation that subtherapeutic doses of selective antibiotic increase the rate of plasmid transfer has important implications for both the design in GMOs and the use or misuse of antibiotics in clinical practice and agriculture.

## Persistence of GMOs and Their Recombinant DNA in vivo

The ability of GMOs, which may be ingested in a viable form in fermented foods, to persist in the human gastrointestinal tract will greatly determine the extent to which recombinant DNA may be transferred to the human gut microflora.

Similarly the stability of recombinant DNA within GMOs in vivo is of major interest when designing GMOs for use in food or in the human gut itself. Stable maintenance of recombinant DNA will greatly determine the ability of a GMO to effectively carry out the task for which it was designed. On the other hand, a certain degree of recombinant DNA instability may limit the risks posed by recombinant DNA persisting in non-target ecosystems.

Gruzza *et al.* investigated the colonization potential and recombinant DNA stability of genetically modified L. lactis strains in germ-free mice. All L. lactis constructs reverted to plasmid-free derivatives in the digestive tract of anexic mice and plasmid-free derivatives became dominant over plasmid-bearing parental strains. Two high copy number, non-self transmissible plasmids exerted strong ecological disadvantage on plasmid-bearing L. lactis strains and were quickly lost from the digestive tracts of the anexic mice.

Transfer of the conjugative plasmid pIL205 was observed between an Ent. faecalis strain colonizing the digestive tract of anexic mice and an L. lactis donor strain. Both the L. lactis donor strain and Ent. faecalis (pIL205) transconjugants were lost from the anexic mice 10 days after dosing with the donor strain. Thus it appears that carriage of recombinant DNA on non-conjugative plasmids and indeed conjugative plasmid DNA, imposed considerable ecological disadvantages on plasmid-bearing strains in anexic animals. Schlundt *et al.* investigated transfer of pAMα1 between L. lactis donor and recipient strains in gnotobiotic and conventional rats. In the conventional animals, both donor and recipient strain were quickly lost from the system and no transconjugants were recovered.

In gnotobiotic animals both donor and recipient strain colonized the gastrointestinal tract. Tranconjugants were recovered within a few days of dosing and maintained at between 103 and 105 CFU less than the recipient strain in faecal samples. Upon sampling of luminal contents of the jejunum, caecum and colon it was observed that carriage of the plasmid seemed to endow a competitive advantage to transconjugants in the small intestine but not in the caecum.

The population of transconjugants was maintained at about 104 CFU/g throughout the intestine but numbers of recipient strain varied from between 104 and 105 CFU/g jejunum contents to 108-109 CFU/g caecal and colonic contents. Clearly the ability of plasmid DNA either to impose ecological advantage or disadvantage on the bacterial host will depend on the plasmid and bacterium involved, and may very well vary considerably between different environmental habitats or microhabitats.

## Effect of Microbial Interactions on DNA Transfer

The human gut microflora exerts considerable colonization resistance on extraneous microorganisms entering the colonic environment. Limiting the persistence of a GMO in a given environment will also limit its ability to undergo DNA transfer with the microflora. Gruzza *et al.* investigated the ability of members of the human gut microflora to limit the colonization ability of and DNA transfer from, genetically modified Lactococcus lactis strains, similar to those likely to be used in the food industry.

L. lactis strains bearing the conjugative plasmid pIL205 and/or a non-conjugative but a pIL205 mobilizable plasmid pIL253 were used. Gnotobiotic mice were associated with four strictly anaerobic bacteria commonly found in the human gut microflora, i.e. Bacteroides sp., Bifidobacterium sp., Peptostreptococcus sp. and Ent. faecalis.

No transfer of plasmid DNA was observed from L. lactis donor strain to Bacteroides sp., Bifidobacterium sp. or the Peptostreptococcus sp. However, upon addition of Ent. faecalis to the polyassociated mice, Ent. faecalis transconjugants were observed despite the competitive exclusion imposed on the L. lactis donor strain by the colonizing Ent. faecalis strain. Gene transfer from the L. lactis strains was also investigated in HFA mice. HFA mice were dosed with L. lactis donor strain, bearing pIL205 and pIL253, seven days after introduction of the human faecal microflora.

From an initial inoculum of 8 × 108 CFU, no L. lactis donor cells were recovered from the faecal samples 24 hours after dosing (detection limit 102 CFU/g faeces). However, 21 hours after introduction of the donor strain, transconjugant facultatively anaerobic streptococci were recovered resistant to antibiotic markers borne on each of the two plasmids.

The number of transconjugants increased until day 15 after dosing. Thus, although rather quickly eliminated from the HFA mice, the L. lactis donor strain was able to transfer its plasmid DNA (both the conjugative pIL205 and the non-conjugative but mobilizable pIL253) to members of the human gut microflora in HFA mice.

Brockmann *et al.* investigated the transfer of plasmid pAMâ1 and pLMP1 (bearing a genetically modified proteinase gene) from L. lactis donor strains to a proteinase-deficient recipient strain in gnotobiotic and conventional rats.

No transfer of the proteinase gene to the recipient L. lactis strain was observed in anexic or conventional rats. Plasmid pAMâ1 was transferred between L. lactis donor and recipient strains in anexic animals but not in conventional rats where both donor and recipient were quickly lost from faecal samples. Transfer of pAMâ1 from the L. lactis donor strain to Ent. faecalis (or a close relative) was observed in one conventional rat.

Using an in vitro single stage continuous flow culture (CFC) of human faecal microflora and in vivo human flora-associated (HFA) rat model of the human colonic microflora we have shown that the human gut microflora exerts a considerable barrier to persistence of L. lactis strains in the colonic ecosystem. A plasmid-free L. lactis strain MG1614 was washed out from the in vitro model at a rate greater than expected for non-growing cells.

Similarly, L. lactis MG1614 introduced into HFA rats was eliminated from the HFA rats within eight days. By continuously dosing HFA rats with L. lactis MG1614 in the drinking water we succeeded in maintaining a stable population of L. lactis MG1614 in the rats. It was hoped that this strain implanted in the microflora of HFA rats would act as a recipient in plasmid transfer studies with a closely related L. lactis strain.

Transconjugants were maintained at between 104 and 105 CFU/g faeces for 27 days. Thus despite the fact the human gut microflora exerts such a drastic barrier effect on the persistence of L. lactis strains in models of the human colonic microflora, transfer of plasmid DNA from an ingested L. lactis donor strain to Enterococcus spp. present in the microflora was observed.

## Design of GMOs and Colonization of Gut by GMOs

GMOs considered for release into the environment in a viable form will employ several measures to limit the ability of recombinant DNA to be transferred to indigenous bacteria in the target environment. Recombinant DNA may be borne on non-conjugative vectors, on the chromosome of the host strain or by the use of suicidal genetic elements.

However, inbuilt safeguards which prevent the transfer of recombinant DNA under labouratory conditions cannot be relied upon to prevent transfer of recombinant DNA in the natural environment. Thus studies employing in vivo models of the human gastrointestinal tract have been conducted with such GMOs in order to determine their ability to persist in the digestive tract and transfer DNA. Gruzza *et al.* investigated the ability of genetically modified L. lactis strains to colonize gnotobiotic mice.

The genetically modified L. lactis strains were constructed with antibiotic-resistance marker genes inserted on different replicons, chloramphenicol resistance was inserted into a self-transmissible plasmid pIL205, erythromycin resistance was inserted into non-self transmissible plasmids pIL252 (low copy number) and pIL253 (high copy number) and a second erythromycin-resistance determinant was implanted into the L. lactis chromosome.

All strains carried a naturally occurring pIL9 plasmid encoding the machinery for lactose fermentation. It was observed that anexic mice were colonized with the recombinant strains or their plasmid-free derivatives, with plasmid-free derivatives becoming dominant in all cases. Plasmids pIL9 and pIL205 were lost from donor strains but plasmid-bearing parental strains and plasmid-free derivatives were maintained at co-dominant levels, suggesting equilibrium between plasmid loss and conjugation in the mouse gastrointestinal tract.

L. lactis strains bearing the low-copy number non-self transmissible plasmid pIL252 rapidly lost their plasmid complement in the gut of anexic mice, probably due to high segregational instability of this plasmid observed in vitro. L. lactis strains bearing pIL253, the high-copy number, non-self transmissible plasmid colonized the digestive tract of anexic mice but only became dominant when plasmid pIL205 was not present in the same cell, indicating a degree of incompatibility between the two plasmids both plasmids belong to the same incompatibility group. The erythromycin-resistance determinant borne chromosomally was maintained stably in the digestive tract of anexic mice. The authors concluded that in vitro determinations of recombinant DNA stability may highlight if a particular insert is extremely unstable but does not give reliable data on the in vivo stability of the construct.

This, highlights the need for in vivo studies both on the stability of recombinant DNA for efficient GMO function and the effect instability has on the persistence of recombinant DNA in the digestive tract, which is a major factor governing the extent to which DNA transfer between ingested GMOs and members of the human gut microflora occurs.

Duval-Iflah *et al.* investigated the mobilization of pBR-derived recombinant plasmids from donor E. coli K12 strains to E. coli strains of human origin in vitro and in the digestive tract of anexic and gnotobiotic mice associated with the human gut microflora. The genetically modified plasmids used were designed to be non-conjugative (mob-, tra-) and either mobilizable or non-mobilizable. E. coli strains isolated from human faeces were screened for their ability to mobilize the oriT- and oriT+ plasmids.

In anexic mice, associated with donor and recipient E. coli strains, mobilization of an oriT+ pBR derivative by trans-complementation with the gene products of tra (encoded by conjugative plasmid R388) and mob was observed. Such mobilization was only observed sporadically with one E. coli strain of human origin in di-associated mice and not at all in HFA mice. It was determined that about 50% of E. coli strains of human origin were able to mobilize oriT- pBR derivatives in in vitro matings but not in anexic mice or HFA animals.

## Dietary Factors Affecting DNA Transfer

The ecological significance of the metabolic burden imposed on a bacterium by plasmid carriage will be greatly determined by the availability of substrates and nutrients to the cell. Similarly, densities of donor and recipient cells have been shown to affect conjugation frequencies in vitro. In our studies, transfer of plasmid pAMâ1 from an L. lactis donor strain to an Enterococcus spp. recipient strain naturally present in the microflora of HFA rats was found to be strongly affected by dose of donor strain employed.

Significant numbers of Enterococcus spp. (pAMâ1) transconjugants were recovered only at a donor cell concentration of 109 CFU/ml or above. The diet fed to the HFA rats also had an effect on transfer of pAMα1 from the L. lactis donor strain to the Enterococcus spp. in vivo.

A high-fat, low-fibre synthetic diet (similar to human high-fat diets) reduced the numbers of Enterococcus spp. (pAMα1) transconjugants recovered from the faeces of HFA rats compared with the commercial rodent chow (higher levels of complex carbohydrate and fibre).

The presence of increased substrate availability in the caecum of HFA rats fed the rodent chow or increased surfaces provided by the higher fibre content of the rodent chow may be involved in the higher numbers of Enterococcus spp. (pAMα1) transconjugants recovered from HFA rats fed this diet. Duval-Iflah *et al.* looked at the effect of milk fermented with Lactobacillus bulgaricus and Streptococcus thermophilus on the transfer and mobilization of plasmids between E. coli strains in the gut of anexic mice.

Plasmid transfer studies were conducted in gnotobiotic mice associated with E. coli K12 donor strain and an E. coli recipient strain PG1 of human origin. The affect of milk fermented with L. bulgaricus, S. thermophilus or with both lactic acid bacteria on transfer of plasmid between the E. coli strains was determined.

Fermented milk consumption had little effect on the populations of R388 or pUB2380 transconjugants in the anexic mice but the milk fermented with both LAB resulted in a slight increase in E. coli PG1 (R388) transconjugants. Long-term consumption of milk fermented with both LAB strains resulted in inhibition of occurrence and maintenance of E. coli PG1 (pCE325) transconjugants in the anexic mice. L. bulgaricus-fermented milk resulted in a reversible reduction in the numbers of E. coli PG1 (pCE325) transconjugants in mouse faecal samples.

When L. bulgaricus and S. thermophilus were grown in BHI broth, the transconjugant lowering capabilities observed with the LAB plus fermented milk were reversed. Administration of L. bulgaricus and the combination of both strains grown in BHI broth increased the population of E. coli PG1 (pCE325) transconjugants recovered from mouse faecal samples. Thus, the authors concluded that probiotic LAB have an effect on the formation and subsequent maintenance of transconjugants in the gastrointestinal tract of anexic mice, and that this effect was dependent on whether the LAB were administered in fermented milk or labouratory growth media, indicating that bacterial metabolic end products as well as viable LAB may be involved.

Clearly much work remains to be done to elucidate the role played by specific environmental parameters, such as diet, ratio of donor and recipient cells, activity of donor and recipient and antagonistic effects of the human gut microflora on plasmid transfer and persistence of GMOs which may be ingested in a viable form in fermented foods.

Genetic engineering is now being employed by the food industry to improve the performance of microorganisms used in the production of fermented foods. Concern has been expressed over the exposure of consumers to genetically modified microorganisms which may be ingested in a viable form in future food products. Of chief concern is the risk of recombinant DNA transfer from ingested GMOs to members of the human gut microflora. A range of natural DNA transfer mechanisms are available to microorganisms in the environment. Chief among these are conjugation, transformation and transduction.

Apart from one case of transformation in the oral cavity, most studies examining DNA transfer in the human gastrointestinal microflora have been concerned with conjugation. In vitro and in vivo models of varying degrees of complexity have been employed to simulate the prevailing ecological conditions in the human gastrointestinal microflora. Such studies have demonstrated the transfer of plasmid DNA, both naturally occurring and genetically modified, in the digestive tract.

Some of the factors thought to affect conjugation between bacteria in the gut microflora have also been studied in vitro and in vivo, e.g. metabolic burden of plasmid maintenance, recombinant DNA stability, colonisation ability of GMOs and antagonistic activities of the microflora, presence or absence of antibiotics, cell concentrations of bacteria involved in conjugation and the effect of dietary supplements on DNA transfer events in the human gut microflora.

In order to provide information of direct relevance to risk assessment of the use of genetically modified microorganisms in food, further scientific investigations employing in vivo and in vitro models.

## GENETICAL MODIFIED ORGANISM-DERIVED FOODS

The concept of substantial equivalence was originally developed through discussions at the Organisation for Economic Co-operation and Development, though, to a large extent, these discussions built on previous work done by the World Health Organization (WHO) and the Food and Agriculture Organization. In 1991, a Joint FAO/WHO Consultation had concluded that the evaluation of a food derived through modern biotechnology should consider both food safety and nutritional value using similar conventional food products as a standard and taking into account the processing of the food and its intended use.

In the 1980s, the OECD established a Group of National Experts on Safety in Biotechnology (GNE), which continued to work through to 1993. As an intergovernmental organization, OECD's GNE was attended by delegates nominated by the governments of the OECD member countries, who were, for the most part, representatives of those agencies and ministries with a responsibility for safety in biotechnology. As a result of the work of the GNE, the OECD published a number of important documents relevant to the safety assessment of biotechnology-derived products during this period, which deal with a range of issues, including safety considerations for industrial, agricultural as well as environmental applications of organisms derived by recombinant DNA techniques.

It was recognized at this stage that the safety assessment of an organism derived through recombinant DNA techniques would rely heavily on the knowledge of its parental organism, as well as on an analysis of how the new organism appears to differ from the parent. OECD recommended that the considerable data on environmental and human health effects of living organisms that exists should be used to guide the risk assessment.

By the early 1990s, there had already been large numbers of small-scale field trials of new crop varieties derived through recombinant DNA techniques, and it became clear that new foods derived from these varieties would be marketed during the 1990s. In order to proactively address food safety issues related to these novel foods, in 1990, the GNE established a Working Group on Food Safety and Biotechnology comprising experts, mainly from the ministries and agencies responsible for food safety issues in OECD member countries. The main objective of the Working Group was to elabourate scientific principles and concepts to be used when evaluating the safety of new foods and food components of terrestrial microbial, plant or animal origin.

The Working Group did not consider the safety assessment of food additives, contaminants, processing aids or packaging materials. Nor did it consider environmental safety issues which had been (or were being) addressed by other groups of the GNE. Early in the life of the Working Group, an important recognition was that traditionally, the safety of food for human consumption had been based on the reasonable certainty that no harm will result from intended uses under the anticipated conditions of consumption. Foods prepared and used in traditional ways have usually been considered safe on the basis of long-term experience, even though they may have contained natural toxicants or anti-nutritional substances.

Normally, new varieties of foods or crops have not been subjected to traditional toxicological testing. In fact, where toxicological testing has been applied to whole foods, the results have often been difficult to interpret. Although the OECD's Working Group recognized that modern biotechnology might extend the scope of genetic changes that can be made - and might even broaden the range of the possible sources of foods - it was recognized that this would not inherently lead to foods that are less safe than those developed through conventional techniques.

The Working Group therefore indicated that the evaluation of foods or food components derived through modern biotechnology does not require a fundamental change in established principles, nor does it require a different standard of safety. Realizing the importance of using examples of new foods to identify and demonstrate the applicability of the proposed scientific principles, the Working Group organized a number of meetings and intergovernmental consultations which included case study presentations of novel foods such as enzymes, genetically modified bakers' yeast, mycoprotein and a number of genetically modified varieties of crop species.

Although these case studies were not intended to be formal safety evaluations they were crucial in illustrating important concepts in the safety assessment of novel foods. Therefore, it was through their application in real examples that the concepts and principles were identified. The Working Group proposed a scientific approach to the evaluation of foods derived through modern biotechnology, which is based on a comparison with traditional foods that have a safe history of use. One of the main concepts developed was that of substantial equivalence, and in describing this concept, the Group stressed that this was a principle that had been used in the past (perhaps intuitively) even if it had not been articulated as such.

The concept of substantial equivalence is described as embodying the idea that existing organisms used as food, or as a source of food, can be used as the basis for comparison when assessing the safety of human consumption of a food or food component that has been modified or is new. As previously indicated, in elabourating the concept of substantial equivalence, the OECD Working Group noted that food safety is considered as a reasonable certainty, that no harm will result from intended uses under the anticipated conditions of consumption and that the most practical approach to the determination of safety is to consider whether a genetically modified organism (GMO)-derived food is comparable to an analogous conventional food product.

The concept of substantial equivalence therefore relies on the existing history of safe food use of a conventional food product as a useful consideration in framing the safety assessment of a GMO-derived food by permitting the identification of similarities and differences which can be considered in the assessment. In 1996, participants at an expert FAO/WHO consultation recommended that safety assessment based upon the concept of substantial equivalence be applied in establishing the safety of foods and food components derived from genetically modified organisms.

Assessing substantial equivalence was recognized as not being a safety assessment per se, but a process which establishes that the characteristics and composition of the new GMO-derived food are comparable to those of a familiar, conventional food which has a history of safe consumption. A Joint FAO/WHO Expert Consultation on Foods Derived from Biotechnology was convened in 2000 to address food safety and nutritional questions regarding foods derived from GM plants, including a review of the scientific basis, application and limitations of the concept of substantial equivalence.

It concluded, based on the current application of substantial equivalence and alternative strategies that this concept contributed to a robust safety assessment framework. Moreover, the Consultation noted that substantial equivalence is a concept used to identify similarities and differences between GM food and a comparator with a history of safe food use which subsequently guides the safety assessment process.

While not a safety assessment per se, the substantial equivalence approach allows the structuring of the safety assessment through characterizing the similarities and identifying the differences which can then be the focus of further consideration. This approach has been referred to as a useful safety standard. It therefore permits inference that the new food under consideration will be no less safe than the conventional food under conditions of similar exposure, consumption patterns and processing practices.

Substantial equivalence is therefore clearly not intended to be a measure of absolute safety, but instead recognizes that while demonstrating absolute safety is an impractical goal, demonstrating that there is reasonable assurance that the GMO-derived product under consideration is no less safe than a conventional food product is an achievable goal. Since it's development, the use of the substantial equivalence concept in the safety assessment of GMO-derived foods has been subject to misinterpretation and criticism.

The comparative nature of the substantial equivalence concept without prescribing the extent of the phenotypic and compositional comparisons has led some to criticize the concept as not being measurable, and therefore inappropriate in safety assessment. This and other criticisms relate, in part, to the mistaken perception that the determination of substantial equivalence was the end point of a safety assessment rather than the starting point.

After several OECD countries had gained experience with safety assessment of GMO-derived foods, an OECD workshop examined the effectiveness of the application of substantial equivalence in safety assessment. This workshop concluded that the substantial equivalence approach provides equal or increased assurance of the safety of foods derived from genetically modified plants, as compared with foods derived through conventional methods.

In 2000, the OECD Task Force for the Safety of Novel Foods and Feeds also reviewed the substantial equivalence concept, its interpretation and its application. While noting that the majority of international guidance documents addressing the safety assessment of genetically modified (GM) plants had interpreted substantial equivalence consistently, this Task Force reported that there were differences in how it was applied and that these differences needed to be resolved.

The Codex Ad Hoc Intergovernmental Task Force on Foods Derived from Biotechnology pursued the development of international guidance on the safety assessment of GMO-derived foods. To date, this Task Force has developed proposed draft principles and guidelines for the safety assessment of foods derived from modern biotechnology. These guidelines interpreted the concept of substantial equivalence as a way of structuring the safety assessment, consistent with the FAO/ WHO interpretation.

The 'Draft Principles for the Risk Analysis of Foods Derived from Modern Biotechnology' and 'Draft Guideline for the Conduct of Food Safety Assessment of Foods Derived from Recombinant-DNA Plants' have been forwarded by the Codex Task Force to the 25th Session of the Codex Alimentarius Commission for adoption at Step 8 of the Codex procedure. A proposed draft guideline for the food safety assessment of foods produced using recombinant-DNA organisms continues to be developed. Substantial equivalence: the application

A conclusion from a 1995 WHO workshop on 'Application of the Principles of Substantial Equivalence to the Safety Evaluation of Foods or Food Components from Plants Derived by Modern Biotechnology' summarized the relationship between substantial equivalence and safety assessment as follows: Establishment of substantial equivalence is not a traditional safety assessment in itself, but a dynamic, analytical exercise in the assessment of the relative safety of a new food or food component to an existing food or food component. The application of the substantial equivalence concept is a key step in structuring the safety assessment of GMO-derived foods. Through appropriate analysis, the new food is compared phenotypically and compositionally to its conventional counterpart with a history of safe use. In this compositional comparison, the characteristics, including levels of key nutrients and toxicants, are considered relative to those of the conventional counterpart taking into account the natural variation for such characteristics.

When a genetic modification results in the insertion of a specific trait, the comparative analysis of the new food will identify both the intended effect (inserted trait) and the potential unintended effects of the modification. Further assessment can then focus on new or altered characteristics for which no history of safe use can be established.

One of the important benefits of applying the substantial equivalence concept is that it provides flexibility which can be a powerful tool in terms of food safety assessment. The comparative approach to structuring the safety assessment can be applied at several potential levels along the food continuum (i.e. harvested primary food material or unprocessed food product, individual processed fractions, or final food product or ingredient).

While from a practical point of view, the compositional comparison should typically be applied at the level of the unprocessed food product, the flexibility of the concept permits the determination to be targeted to the most appropriate level based upon the nature of the product under consideration. Where multiple fractions from a single source are destined to different food products, the comparative approach might be targeted at the level of the unprocessed food product in order to permit the safety assessment to apply to all derived fractions (e.g. for soybean, where multiple fractions are used widely in foods, assessment at the level of the seed is appropriate).

However, where a single fraction of a particular raw material is used as human food, then the comparison can focus at the level of the single fraction, thereby simplifying the safety assessment process (e.g. for canola, where the processed oil is the only fraction consumed by humans, safety assessment may appropriately be focused on comparison of the oil composition of the novel variety with traditional canola oil composition).

Application of the substantial equivalence concept in the safety assessment of a GMO-derived food depends on the identification of an appropriate comparator with an acceptable history of safe food use. It also requires that sufficient analytical data be available in the literature or be generated through analysis to permit an effective comparison. These requirements present a key limitation of the substantial equivalence concept since the consideration of similarities can only provide assurance of safety relative to those components assessed for the particular comparator. The choice of the comparator is therefore crucial to the effective application of substantial equivalence in establishing the safety of a GMO-derived food.

An appropriate comparator must have a well-documented history of use. If adverse effects have been associated with the particular food type, specific components of the food which are considered to be causative of those adverse effects should be described and well characterized in order to permit effective comparison. A joint FAO/WHO Expert Consultation on Biotechnology and Food Safety considered the application of substantial equivalence in the safety assessment of GMO-derived foods.

The consultation recommended that applicacation of the substantial equivalence concept entail consideration of the molecular characterization of the new food source; phenotypic characterization of the new food source in comparison to an appropriate comparator already in the food supply; and the compositional analysis of the new food source or the specific food product in comparison to the selected comparator. The guidance further elabourates on compositional comparison by highlighting what information should provide sufficient information to permit effective comparison.

The recommended focus of the compositional comparison is on analytical comparison of those components identified in the food source in question which are nutrients which provide a substantial impact in the overall diets (key nutrients) and toxicologically significant compounds known to be inherently present in the species (key toxicants). In addition, there is recognition that additional components might be identified for analysis based upon the molecular and phenotypic characterization and the nature of the genetic modification. In addition to key nutrients and toxicants, additional parameters may be appropriate for assessing the potential for unintended effects of a genetic modification.

When modifications are directed at metabolic pathways of key macro or micro nutrients, the possibility of an impact on nutritional value is increased, particularly if that food is a major dietary source of the nutrient affected. The potential for unintended effects would be determined, in part, by the nature of the intended alteration (e.g. consideration of fatty acid profile if an enzyme involved in fatty acid metabolism is introduced) and the data from molecular and phenotypic characterization. The consideration of key nutrients and key toxicants in the comparison which is essential to applying a substantial equivalence in safety assessments also introduces another limitation of the concept. The nature of the comparative approach with respect to nutrients limits its universality since the relevance of nutrients in a particular crop are dependent on consumption patterns which might vary from region to region. Where differences in consumption exist for a particular crop, these must be considered in the identification of the key nutrients for assessment.

This is particularly true for crops which form a significant portion in the diet in a particular region. The comparative approach to assessment can be applied in each region, but conclusions for one region will not automatically hold for another region if there are significant differences in consumption patterns and processing practices. Another potential regional limitation is related to the application of the concept of substantial equivalence as opposed to an inherent limitation of the concept per se.

It may be difficult, particularly in developing countries, to apply the concept to assess the safety of foods where adequate nutritional databases are not available for a given population. A safety assessment using the substantial equivalence approach does not demonstrate that a GMO-derived product is identical to its conventional comparator since the compositional comparison does not take into account all components.

However, application of the guidance provides assurance that the comparison has considered those components most likely to be relevant to the safety of the product as it is expected to be consumed in a particular region. Recognizing the importance of there being compositional data available for applying substantial equivalence, the OECD Task Force for the Safety Assessment of Novel Foods and Feeds has focused on the development of science-based consensus documents containing information on the nutrients, anti-nutrients or toxicants, product use and other data relevant to the assessment.

The OECD have published such documents for potatoes, sugar beet, soybean and low erucic acid rapeseed (canola). Such guidance will add to the understanding of appropriate parameters for a comparative assessment of composition while recognizing that additional parameters may be relevant to the safety assessment dependent on differences in consumption pattern. The international guidance developed by the OECD and FAO/WHO has been practically applied to the safety assessment of GMO-derived food products in several countries. In order to facilitate such assessments, specific guidance documents which embrace the substantial equivalence concept have been published. The majorities of these currently address the safety assessment of genetically modified plants and have consistently interpreted the concept of substantial equivalence. Internationally, these guidance documents have been applied to the assessment of a significant number of GMO-derived plant products over a period of more than eight years, demonstrating that the concept of substantial equivalence can be applied effectively in the safety assessment of novel foods.

## RAPESEED OIL (CANOLA) AS A CASE STUDY

The comparative approach and hence the concept of substantial equivalence has been applied to the safety assessment of new and modified foods derived from plants developed using traditional breeding practices over the last 20 years and more recently to those products derived from recombinant DNA technology.

Just as the OECD Working Group utilized case studies, the application of substantial equivalence in structuring the safety assessment of novel foods can be illustrated through discussion of the modifications made to rapeseed oil through traditional and recombinant DNA techniques.

### Developments in Rapeseed Oil (canola)

Rapeseed breeding in Canada began soon after the crop was introduced during World War II. The initial goals of breeding were directed towards improving agronomic characteristics and oil content of rapeseed. Nutritional experiments conducted as early as 1949 indicated that consumption of large amounts of rapeseed oil with high levels of erucic acid could be detrimental to experimental animals.

Concerns about the nutritional safety of rapeseed oil and the potential impact on human health stimulated plant breeders to search for genetically controlled low levels of erucic acid in rapeseed oil. After 10 years of backcrossing and selection to transfer the low erucic acid trait into agronomically adapted cultivars, the first low erucic acid varieties, Brassica napus and B. campestris were released in 1968 and 1971, respectively. Rapeseed meal is used exclusively in Canada as a high-protein feed supplement for livestock and poultry.

Prior to the late 1970s, the use of this oilseed processing byproduct as an animal feed was limited by the presence of glucosinolates in the seed. The low palatability and the adverse effect of glucosinolates due to their anti-thyroid activity led to the development of varieties of rapeseed which have combined low levels of both glucosinolates and erucic acid (also known as 'double low' varieties). Canola breeding programmes in the 1980s and 1990s have produced cultivars with higher yields, increased oil and protein contents, earlier maturity, yellow seeds, reduced green seed and improved disease, insect and herbicide resistance.

(Note: The term canola has been registered and adopted in Canada to describe the oil (seeds, plants) obtained from the cultivars B. napus and B. campestris. In 1986 the definition of canola was amended to refer to B. napus and B. campestris lines containing <2% erucic acid in the oil and <30 mmol/g glucosinolates in the air-dried, oil-free meal.)

## Low-erucic Acid Rapeseed Oil

In 1987 low-erucic acid rapeseed oil (LEAR oil) was given generally recognized as safe (GRAS) status in the United States. Although this product was not produced as a result of recombinant DNA technology, it was considered to be a 'novel food'. The LEAR oil case study illustrated the application of the principles developed by the Working Group (established by the GNE) for assessing the safety of foods by applying the concept of substantial equivalence. The novel trait of LEAR oil was the low erucic acid content when compared with traditional rapeseed oil. In this case precise comparisons between LEAR oil and other vegetable oils could not be made because vegetable oils vary in composition depending upon the variety of plant and the growing conditions.

The concept of substantial equivalence was applied to assess the safety of LEAR oil by comparing the individual fatty acid components to similar components present in other traditional oils including soy, corn, peanut, safflower, olive and sunflower. Except for the low levels of erucic acid, the individual fatty acid components of the LEAR oil were comparable to those similar fatty acids found in common vegetable oils.

Dietary exposure estimates for average and upper limit intakes for LEAR oil used by itself and as a component of blended oil products including shortening, margarine, salad oil and vegetable oil did not raise any safety concerns. The data considered in assessing the nutritional adequacy and digestibility of LEAR oil were similar to those that would be considered for any new oil (i.e. human and animal feeding studies).

Due to concerns regarding the safety of the erucic acid component in both traditional rapeseed and LEAR oils, toxicological studies were an important consideration in the safety assessment. These studies included a review of an extensive toxicological database and a scientific rationale supported by the results of animal feeding studies in several species with a range of vegetable oils. The development of LEAR/canola and the application of the comparative approach to directing its assessment is illustrative. We can further build on this illustration by considering the application of recombinant DNA technology to canola to develop herbicide-tolerant varieties.

In these cases, considerations for the safety of the oil derived from these varieties follow the same approach. In this case the appropriate comparator is now the unmodified canola variety. The database used in establishing the safety of LEAR/canola oil provides an effective tool for the comparison of the composition of the oil derived from the herbicide-tolerant variety under consideration.

Comparison of the oil from new canola varieties developed through recombinant DNA technology with the oil from unmodified canola with the same food use is an effective approach to determine its safety. Consideration is appropriately given to key nutrients (i.e. fatty acids) and key toxicants (i.e. ensuring that the level of erucic acid is sufficiently low for safe consumption). The comparative approach can also be applied in the safety assessment of other compounds in plants such as canola.

The canola meal, which may be used as an animal feed, would require a separate evaluation since measurable amounts of the introduced protein may be present in the meal, unlike in the refined oil. In the case of the meal, the safety of that protein in animal feeding may need to be demonstrated if the protein is new to the feed supply.

If, based on the compositional comparison of the canola meal to an unmodified counterpart, the only difference is the presence of the introduced protein, then the next steps in the safety assessment would focus on determining if that new protein had the potential to be toxic or impact the nutritional quality for feed use.

The successful lowering of erucic acid led to continued interest in the compositional modification of canola oil. For example, plant breeders have used mutagenesis to genetically alter the plant's fatty acid biosynthetic pathways to obtain specialized fatty acid compositions. Canola oil has been developed with the linolenic acid content reduced from approximately 10% to less than 3%. Although high levels of linolenic acid are desirable from a nutritional point of view, they are undesirable in terms of chemical stability.

Other recent developments in canola oil compositional changes include the application of mutagenesis to produce high levels of oleic acid. The resulting high oleic acid-producing cultivar was then crossed to low linolenic cultivars to create a high oleic/low linolenic line. Intentional modifications in the composition of canola oil may lead to the production of canola varieties that are not like other commercial varieties. For example, in the last 10 years the application of recombinant DNA technology has resulted in the production of increased levels of lauric and myristic acids in canola oil.

Similarly, the application of mutagenesis has produced higher levels of oleic acid in canola oil. For these canola varieties the fatty acid profiles and levels will not fall within the ranges defined in the Codex standard for Edible Low Erucic Acid Rapeseed Oil or the Codex draft standard for Named Vegetable Oils (which includes low erucic acid rapeseed oil).

In cases where the fatty acid composition of canola oil has been intentionally modified so that commercial canola varieties cannot be used as a comparator, the fatty acid component(s) will need to be considered on an individual basis using other commonly consumed oils as appropriate for comparison where such fatty acid component(s) are present in similar levels.

Substantial equivalence could be applied at the component level to assess the safety of the oil produced since these fatty acids have a safe history of consumption as a significant component of other edible oils. In addition to fatty acid profiles and levels, modifications may also result in alterations in the chemical structure (e.g. saturation, chain length and triglyceride structure) that may have nutritional consequences or result in changes in digestibility.

Chemically altered fatty acids will need to be evaluated on their merit, which may involve a combination of nutritional and toxicological in vivo and in vitro testing. Laurate canola provides an example of oil that is not compositionally identical to any other food oils, although it shares many similar characteristics.

Unlike other commercial canola varieties which contain no detectable lauric acid, laurate canola produces high levels of lauric acid. Laurate canola also produces lower levels of oleic acid and higher levels of myristate compared with the Codex specifications for those fatty acids in low erucic acid rapeseed oil. Other fatty acids in laurate canola, including palmitic, palmitoleic, stearic, linoleic, linolenic, gadoleic, eicosadienoic, behenic and lignoceric, are similar to those levels found in commercial canola varieties.

Levels of the naturally occurring toxicant erucic acid are very low. Substitution of laurate canola for coconut and palm kernel oils does not raise any safety concerns for the intended uses because the major components, laurate and myristate, are identical. The safety assessment of novel foods based upon the concept of substantial equivalence relies on comparison with the conventional foods with a long history of safe use. The comparator can be the species itself, a product derived from that food source or a similar component from a different species (i.e. fatty acids in vegetable oils). The food safety issues for organisms that have been genetically modified are of the same nature as those that may occur through other ways of genetic modification, such as traditional breeding.

These include potential food safety concerns such as toxicity or allergenicity. Once the safety assessment is completed it is reasonable to assume that the novel food does not pose a risk different to those of traditional foods that have been safely part of the diet for many years. The term 'substantial equivalence' and its application in safety assessment has been criticized or challenged as lacking a clear definition and therefore being imprecise. However, when considered as a concept, the lack of a measurable definition does not restrict its application. Stated most simply, substantial equivalence encourages investigators to compare a product which they have to assess with one with which they are already familiar.

The application of this concept to the assessment of foods derived from GMOs is therefore intended to permit the creation of a linkage between the new GMO-derived food and a familiar, conventional food in terms of their characteristics and composition. Over the past eight years, the concept of substantial equivalence has been consistently interpreted and practically applied by numerous countries assessing the safety of foods derived from genetic modification.

It has been demonstrated that the concept of substantial equivalence can be applied effectively in the safety assessment of those genetically modified foods developed for commercialization to date. It is noteworthy that the Codex Ad Hoc Intergovernmental Task Force on Foods Derived from Biotechnology has supported the comparative approach as a key element in structuring the safety assessment of GMO-derived foods, signifying that there is significant international consensus emerging on the application of substantial equivalence.

While not yet a formal consensus, the agreement of the Codex Task Force to forward the 'Draft Principles for the Risk Analysis of Foods Derived from Modern Biotechnology' and 'Draft Guideline for the Conduct of Food safety Assessment of Foods Derived from Recombinant-DNA Plants' to the Codex Alimentarious Commission for adoption signifies emerging consensus.

## Chapter 9

# Allergenicity of Foods

Immunologically mediated adverse reactions to components normally present in foods, and generally referred to as 'food allergens', are a significant health concern for both the pediatric and the adult population at risk. It is thus necessary to develop rational approaches to prevent or limit the spread of food allergens during a time when it has become technically possible to alter the genetic composition of foods derived from both plant and animal sources.

To complete the argument that genetically modified foods must be considered for allergenic potential, refer to the identification of Brazil nut allergen in transgenic soybean or the ability of Bacillus thuringiensis pesticides to induce the production of specific IgE. The general pathway to protection appears clear: attempt to avoid the transfer of known significant allergens into other foods, i.e. try not to create new significantly allergenic foods.

But are the issues this clear, and what are the difficulties and limitations in this approach? To understand the depth and breadth of this problem, it is first necessary to examine the spectrum and prevalence of food allergies, and where reasonable evidence exists as to the etiologic agents within the foods responsible. Given this database, it then becomes possible to examine what approaches may be used to reduce risk - and the limitations of these strategies.

It is clear that a diversity of adverse food reactions exist - some of which are on an immunologic basis and are thus referred to as 'food allergies'. Following the classification of adverse food reactions as adopted by the European Academy of Allergy and Clinical Immunology, such reactions may be divided into 'toxic' and 'non-toxic' groupings. Toxic reactions are defined as those that may develop in anyone who ingests a sufficient dose. Non-toxic reactions rely on person-to-person susceptibilities.

These reactions may have at their basis either immune or non-immune mechanisms. Non-toxic reactions include those referred to as allergies or hypersensitivities. Adverse reactions from non-immune mechanisms, often referred to as intolerances, may be attributed to pharmacologic properties of the food and/or unique susceptibilities in the individuals affected.

## DEMOGRAPHICS

Prevalence data relating to food allergies are somewhat limited. In a study of Danish infants, the prevalence of cow's milk allergy was found to be 2.2%. In another study, newborns were followed through to their third birthday. Approximately 4% were thought to have IgE-mediated food allergies. Similar studies of food allergy are unusual in adult populations.

One study in the Netherlands, which was based on questionnaires, clinical follow-up and double-blind placebo-controlled food challenge, estimated the prevalence of food allergy and intolerance together to be 2.4%. The prevalence of peanut and tree nut allergy in the US as determined in a random digit dial telephone survey was found to be 1.1% of the general population. Such data gives credence to the conclusion that food allergies are a major health issue worldwide.

## TYPES OF REACTIONS

### Immediate Reaction

IgE-mediated immediate reactions are the basis for the majority of allergic reactions to food, and may result in death from anaphylaxis. Thus, much of the concern that surrounds food allergies and their consequences relate to patients in this clinical category. These responses follow the release of chemical mediators of inflammation from mast cells and basophils following the interaction between food-specific IgE and a specific food allergen on the surfaces of these effector cells. Patients with food allergies appear to represent a subgroup of atopic individuals who have allergies in general and more difficulty regulating IgE levels in response to environmental antigens.

Immediate hypersensitivity reactions to food antigens are evidenced by a spectrum of clinical findings from eczema to anaphylaxis. The oropharynx is the initial site of exposure to food antigens. Edema and pruritus of the lips, oral mucosa and pharynx may be reported as the food contacts the mucosal surfaces. The term oral allergy syndrome is applied to the clinical situation dominated by oropharyngeal symptoms. This is most commonly associated with the ingestion of fruits or vegetables. It has been suggested that the observation that such foods as apple, which cross-reacts with birch, or watermelon and cantaloupe which cross-react with ragweed, relates to a relationship between the oral allergy syndrome and allergic rhinitis. Entry of food into the upper gastrointestinal tract may result in nausea, cramping, pain, abdominal distension, vomiting, flatulence and diarrhea. Symptoms of gastrointestinal involvement may be the only expression of food hypersensitivity, but this is unusual. Food allergy is usually also expressed in one or more extra-intestinal target tissues. The skin is a common target organ. Reactions include acute urticaria, acute angioedema, atopic dermatitis and less frequently chronic urticaria. It has been estimated that clinically significant food hypersensitivity exists in approximately a third of children with atopic dermatitis.

Asthma and rhinitis secondary to food hypersensitivity are more common in children than in adults. Systemic anaphylaxis associated with allergy to ingested foods generally occurs within 1-30 minutes after ingestion of the offending food. The first anaphylactic episode may be unexpected or may be preceded by prior symptoms, such as abdominal discomfort or urticaria, on previous exposure to the food. Anaphylaxis may include tongue itching and swelling, palatal itching, throat itching and tightness, wheezing and cyanosis, chest pain, urticaria, angioedema, abdominal pain, vomiting, diarrhea, hypotension and shock.

Severe life-threatening reactions are most often associated with the ingestion of peanuts, nuts and seeds. Fatal reactions may progress rapidly or begin with mild symptoms and then evolve to cardiorespiratory arrest and shock over hours. Systemic anaphylaxis has also been reported after ingestion of food followed by exercise. Labouratory procedures used to help diagnosis of immediate food hypersensitivity involve the identification of antigen-specific IgE to allergens within suspected foods. In the case of skin testing, the IgE examined is fixed to skin mast cells. In vitro tests identify antigen-specific IgE in serum.

Allergic reactions to foods are unusual in the face of negative tests (false negatives). Patients should never be advised that they are allergic to certain foods solely on the basis of positive tests. This is because tests may also be positive to foods in the absence of symptomatic food allergy (false-positive tests). In cases of the oral allergy syndrome, the use of extracts of fresh fruits and vegetables is often necessary to exclude IgE-mediated food hypersensitivity.

The evaluation of an IgE-mediated food allergy utilizes the medical history, physical examination and relevant labouratory studies. The diagnosis in some instances must be confirmed by blinded food challenge. When evaluating an adverse reaction to food, a number of other diseases, anatomic defects and reactions to additives, toxins and contaminants that may in some way mimic an allergic reaction must be considered and eliminated. Enzyme deficiencies both mimic or may complicate gastrointestinal inflammatory diseases. Abdominal cramping, bloating, and diarrhea accompany the ingestion of milk and milk products in individuals with lactase deficiency.

Deficiencies such as galactose-4-epimerase (galactosemia) may be diagnosed in infancy as a result of vomiting and diarrhea after milk ingestion. Cystic fibrosis may initially be confused with a food-induced malabsorption syndrome because of associated pancreatic enzyme deficiency. Chronic cough and wheezing secondary to aspiration may occur with hiatal hernia, pyloric stenosis and an H-type tracheoesophageal fistula. Overfeeding and chalasia are estimated to occur in up to 50% of newborns and result in vomiting associated with feeding. Abdominal pain following meals may be due to peptic ulcer disease or cholelithiasis.

Drugs, dyes, additives, bacteria and bacterial products may be present in foods. Tartrazine yellow is a rare cause of hives. Sulfiting agents used to reduce spoilage of foods, to inhibit undesirable microorganisms during fermentation, to sanitize food containers and to prevent oxidative discolouration of foods induce problems reminiscent of allergic diseases including bronchospasm.

Monosodium glutamate in sufficient quantity (usually greater than 6 g) may lead to a transient syndrome consisting of warmth or burning sensation over the head and shoulders, stiffness or tightness, extremity weakness, pressure, tingling, headache, light-headedness and gastric discomfort occurring approximately 15 minutes after ingestion. Toxins may induce signs and symptoms resembling allergic reactions.

In scrombroid poisoning, ingestion of fish containing high levels of histamine is followed by symptoms including diffuse erythema and headaches. Scrombroid fish commonly implicated include tuna, skipjack and mackerel. Ciguatera poisoning is seen especially in the Caribbean and Pacific islands.

Symptoms include tingling of the lips, tongue and throat, followed by nausea, vomiting, diarrhea, headache, chills and myalgias. Paralytic shellfish poisoning is caused by ingestion of bivalve mollusks contaminated with dinoflagellates of the genus Gonyaulax which produce neurotoxins. Amnesic shellfish poisoning is an acute illness characterized by gastrointestinal symptoms, seizures, coma, disorientation and loss of memory following the ingestion of mussels. This disease is due to domoic acid, a potent neurotoxin produced by the bloom of the pennate phytoplanktonic diatom Nitzia pungens. The only proven therapy for food allergy is strict elimination of the offending allergens. Severe elimination diets should only be instituted with nutritional guidance. Patients and parents must learn to read and understand food labels.

No appropriately designed trial has demonstrated clear efficacy and practicality for the use of prophylactic medications, injection immunotherapy or oral desensitization in the prevention of allergic reactions to foods. Food hypersensitivity reactions occurring in infants may be delayed, and in some cases avoided by breast feeding. Many infants eventually become tolerant to the foods provoking reactions. In prospective studies of adverse food reactions in infants, 80-87% of confirmed symptoms were no longer observed by three years of age. Older children and even some adults may lose their sensitivity if the responsible food allergen is completely eliminated from the diet.

Thus, after one or two years of allergen avoidance, up to one-third of children and adults lose their clinical sensitivity. Patients with an allergy to peanut, tree nut, fish or shellfish rarely lose their clinical reactivity. Loss of sensitivity correlates with allergen avoidance, but whether reintroduction and repeated exposure to the same allergen will cause the sensitivity to reappear is unknown.

A patient may sometimes inadvertently consume a food to which he or she is sensitive. Treatment for a specific symptom which results from inadvertent exposure is the same as that employed when other factors provoke symptoms. Thus, laryngeal or pulmonary symptoms following an inadvertent food exposure should be treated immediately with epinephrine or bronchodilator therapy or both.

The treatment of food-induced anaphylaxis is essentially the same as that for anaphylaxis due to a medication or insect sting. A patient with potential anaphylactic reactivity should be taught how to self-administer epinephrine, and have an epinephrine-containing syringe and an antihistamine available at all times. For children, daycare centres and schools should have a list of emergency numbers with backups to be called.

It should be remembered that a patient may exhibit only mild symptoms in the first few minutes after ingesting a food to which he or she is allergic, but this may be followed 10-60 minutes later by hypotension and other severe problems. Following self-medication for systemic reactions, the patient should immediately seek medical attention. All patients with IgE-mediated food allergy should be warned about the possibility of developing a severe anaphylactic reaction and should be educated in the appropriate treatment measures to be taken in case of an accidental ingestion.

## DELAYED REACTION

## FOOD PROTEIN-INDUCED GASTROENTEROPATHY

Food protein-mediated gastroenteropathy appears almost exclusively to be a disease of infants and children. Its basis is a hypersensitivity reaction to a food protein which results in damage to the intestinal mucosa. Food protein gastroenteropathy may be induced by cow's milk protein, soy protein, egg, fish, rice or chicken. The basis of food protein-mediated gastroenteropathy is unknown, although a cell-mediated delayed hypersensitivity mechanism has been suggested.

Total serum IgE is often normal, and IgE specific to the inciting protein is usually absent in affected children. The predominant symptoms of food protein-mediated gastroenteropathy are vomiting, diarrhea, malabsorption and gross or occult stool blood loss. Symptoms usually appear within weeks of introducing cow's milk or other proteins into the diet. Carbohydrate malabsorption may be present secondary to an intestinal mucosal injury. Eosinophilia may be present in some patients. The diagnosis of protein hypersensitivity requires the demonstration of a relationship between ingestion of a given protein and symptomatology consistent with intestinal damage.

In children exhibiting diarrhea and malabsorption, biopsy of the small intestine usually reveals villous atrophy, primarily in the jejunum. In some children only the colon is involved, manifesting as colitis. Treatment is dependent on removal of the sensitizing protein from the child's diet. An elemental formula or formulation or total parenteral nutritional may be required to allow for repair and return of normal function of the gastrointestinal mucosa. Reintroduction of the offending protein may be attempted, since protein-mediated gastroenteropathy typically resolves by 18-24 months of age.

### Eosinophilic Gastroenteritis

Eosinophilic gastroenteritis is characterized by eosinophilic infiltration of the gastrointestinal wall, peripheral eosinophilia and gastrointestinal symptoms. The sites of involvement include the esophagus, stomach, small intestine, colon and, rarely, extra-intestinal organs. Approximately half of the cases have allergic features and may be related to food allergy.

It may occur at any age, although the peak age of onset is in the third decade. Most of those with food-dependent disease are under the age of 20 years. The cause of food-induced eosinophilic gastroenteritis appears to be related to an IgE-dependent, mast cell-mediated mechanism. Atopic diseases, such as eczema, allergic rhinitis, bronchial asthma, urticaria and a positive family background for allergy are common. Many patients also have peripheral eosinophilia, an elevated serum IgE level and positive RASTs for specific IgE antibodies to food antigens. All patients with eosinophilic gastroenteritis, however, are not atopic, and all cases cannot be explained by food allergy. Only approximately half of the patients with eosinophilic gastroenteritis have findings consistent with atopy. Many patients show no personal or family history of allergy, any positive skin tests for food allergens, no elevation in serum IgE, and no adverse reactions to foods.

Even in patients who have suspected food allergies, sequential withdrawal of various food substances may fail to provide amelioration of symptoms, and there may be a poor correlation between results of skin tests to specific food antigens and the results of an elimination diet. In addition, most patients show no abnormality following extensive immunologic studies, including serum IgE, IgM and IgA levels; complement levels; lymphocyte quantification; and lymphocyte responses to non-specific mitogens.

Patients with eosinophilic gastroenteritis experience nausea with vomiting, abdominal pain, diarrhea, steatorrhea and either weight loss in the adult or growth failure in the child. Radiologic findings may reveal mucosal edema and nodularity of the folds of the small bowel. Muscular disease can precipitate small bowel obstruction. Significant mucosal disease typically presents with evidence of iron-deficiency anemia, hypoalbuminemia and hypogammaglobulinemia secondary to a protein-losing gastroenteropathy, steatorrhea and an eosinophilic leukocytosis. The mucosal form of eosinophilic gastroenteritis may be confused with celiac disease, regional enteritis, neoplasms (lymphoma), polyarteritis nodosa, parasites, the hypereosinophilic syndrome or another protein-losing gastroenteropathy.

The diagnosis of eosinophilic gastroenteritis is established with a gastrointestial biopsy demonstrating an eosinophilic infiltration of the gastrointestinal wall. Treatment of eosinophilic gastroenteritis is often unsatisfactory. Food hypersensitivity should be ruled out as the precipitating or exacerbating factor. A trial diet that eliminates multiple food antigens from the diet in a systematic order may be performed when no clear history or evidence of IgE-mediated hypersensitivity is elicited. In patients with eosinophilic gastroenteritis and food sensitivity, the number of foods involved may preclude the long-term use of an elimination diet for symptom control. Both patients who responded poorly to dietary restrictions and those without evidence of food hypersensitivity may require oral steroid therapy.

### Gluten-sensitive Enteropathy (celiac disease)

Gluten-sensitive enteropathy is a mucosal disease of the small intestine caused by the alcohol-soluble portion of gluten (gliadin) in susceptible individuals. The HLA-DR3DQ2 haplotype has the strongest disease associations. Symptom onset typically occurs 6-12 months after introduction of gluten into the diet. Complaints include intermittent diarrhea, abdominal pain and irritability. Extensive mucosal injury may result in malabsorption with a clinical picture including steatorrhea, peripheral edema from protein loss, anemia, bleeding diathesis, and tetany and growth failure.

An increase in the incidence of gastrointestinal lymphoma is reported. The acute reaction of the intestinal mucosa consists of edema, an increase in vascular permeability, and an eosinophil and neutrophil infiltration. Gliadin has been separated by gel electrophoresis into four proline and glutamine-rich fractions, each of which precipitates small bowel injury in vitro. Blunting of the mucosal surface, villous atrophy and a dense infiltration of the lamina propria with plasma cells, B cells and T cells is observed in chronic disease.

The diagnosis is dependent on demonstrating biopsy evidence of small intestinal mucosa injury upon gluten challenge. Treatment is directed at elimination of gluten from the diet. Wheat, barley, rye and oats contain gluten. Improvement in symptoms is seen as soon as two weeks after the institution of a gluten-free diet.

Histologic improvement may take 2-3 months. Growth usually returns to normal once the small intestinal mucosa has healed. Strict gastrointestinal rest, and in some instances the use of steroids, is necessary to suppress diarrhea when inflammation is severe.

### Dermatitis Herpetiformis

Dermatitis herpetiformis is a chronic papulovesicular skin disorder, often associated with asymptomatic gluten-sensitive enteropathy. The histologic appearance of skin lesions is one of a granulocytic infiltration at the dermoepidermal junction associated with edema and blister formation. Granular IgA deposits with associated J chains are found in the papillary dermis. Complement-mediated injury is implicated. The histology of intestinal lesions is similar to celiac disease though usually less severe.

Skin lesions are symmetrically distributed on extensor surfaces of elbows, knees and buttocks. Most patients have little or no gastrointestinal complaints. The diagnosis rests on the typical appearance of skin lesions and histologic findings of IgA deposits in the perilesional or uninvolved skin. Fifteen per cent of patients will have a normal small intestinal mucosa on histologic evaluation. Treatment consists of the removal of gluten from the diet and the use of dapsone or sulfapyridine.

### Respiratory Disease

Asthma due to ingested foods and as part of a systemic IgE-mediated reaction is well recognized. Asthma induced by inhaled aerosolized food components, and with isolated organ involvement also occurs. It is encountered most frequently as 'occupational asthma'. Asthma defined as reversible airway obstruction must be separated from hypersensitivity pneumonitis (HP).

HP is characterized by a diffuse, predominantly mononuclear cell infiltrate in the lung parenchyma, which may be followed by fibrosis. This disease also occurs after exposure to organic dusts, is principally an occupational disease, and is not IgE-mediated. Similar foods (aerosolized during cooking or grinding) may cause both asthma or HP. Therapy is directed at identifying the etiologic agent and practicing avoidance. In some cases this is as direct as improving ventilation. Failure to identify HP in a timely fashion may result in permanent pulmonary function abnormalities.

## CONTACT REACTIONS

Contact dermatitis consists of a rash resulting from a substance touching the skin. Such reactions may be irritant in nature, result from IgE-dependent allergic mechanisms, or be related to photoallergic and delayed hypersensitivity (T cell-mediated eczematous disease). Contact urticaria is a term specifically related to skin disease due to IgE-dependent, immediate, mast cell-related mechanisms.

Occupational eczema has been reported in association with a number of foods including fish, meat, vegetables, crustaceae, celery and spices. Inciting substances are identified by history and patch testing. Treatment is avoidance.

## ALLERGENS

The specificity of IgE-mediated reactions as defined by the identification of allergenspecific IgE, the demonstration that allergens interacting with allergen-specific IgEs on the surface of mast cells and basophils initiate inflammatory reactions, and the association of these observations provoked in vivo in those with immediate reactions to foods to clinical disease, focused attempts at risk protection on such reactions. The evidence is thus persuasive that 'food allergens' as defined by specific IgEs induce the immediate clinical reactions to foods including anaphylaxis, asthma, hives, angioedema and some contact reactions. Indeed, this approach covers all of the immediate, the immediate component of allergic eosinophilic gastroenteritis, and occupational asthma. Unfortunately, the components in foods responsible for food-induced enterocolitic syndromes are poorly characterized, and thus the strategy of preventing spread of known antigens cannot be relied on to protect children with these disorders, although they constitute a small subset of those that experience adverse reactions to foods.

Fortunately, the components within certain foods that provoke celiac disease and dermatitis herpetiformis (gliadins) are known. Advisory groups have uniformly recommended that gliadins not be transferred during genetic modification of foods, seeming to place this issue at rest. Thus, the focus in protection has been on allergens and their characteristics. And while the general focus has been on food allergens, there is general consensus that inhalant allergens, usually not consumed to any degree, should also be considered in protection approaches, even though they are often not in the edible portion of the plant, and knowing that allergens in fruits and vegetables causing the oral allergy syndrome may cross-react with certain inhalant allergens.

Food allergens in general are water-soluble glycoproteins that range in size from 10 to 60 kDa. For poorly understood reasons, they tend to be stable to treatment with heat, acid and proteases. It has been suggested that this is indirect evidence for the conclusion that the allergenic potential of food allergens as defined by IgE binding (B cell epitope) lies in linear, rather than conformational (continuous or discontinuous) epitopes. This is in apposition to the general thought that T cell epitopes are linear, whereas B cell epitopes are conformational.

Allergenic foods for the most part contain multiple significant allergens. Major allergens are usually defined by the observation that more than 50% of patients sensitive to that food react with these proteins. While virtually any food may cause an allergic reaction in someone, somewhere, and extensive surveys of such occurrences have been published, a relatively small number of foods appear to cause over 90% of all reported immediate reactions in both children and adults.

These allergens are named according to the accepted taxonomic designation, using the first three letters of the genus and the first letter of the species with an Arabic number assigned in the order of their designation (the same number is used to designate homologous antigens). In infants less than six months of age, the majority of allergic reactions are due to milk or soy. In adults, the most common food allergens are peanuts, tree nuts, crustaceae, fish and egg. It has been recommended that sesame seed and wheat be added to this list. It has also been suggested that members of the Prunoideae subfamily (peach, plum, apricot, cherry, almond), celery and rice may have to be added to the list.

## RISK PROTECTION

Given that protection strategies focus on the avoidance of the creation of new and significant allergenic foods as defined by alterations or additions to the genome of a plant or animal that would result in the synthesis by that organism of a newly expressed allergen, there are at least three major considerations. First, how to avoid transferring a known major allergen from one source to another.

Second, what strategies may be employed to identify whether a gene selected for transfer codes for a protein with significant allergenic potential where there is no previous significant population exposure to that protein? Third, how to monitor modified foods to see if modifications led to the expression of new allergens, or upregulated the expression of existing allergens.

Individuals and groups reviewing these issues have generally identified four technical approaches that may be used to determine if a modified food may be or is enhanced in its allergenic potential. It is also clear that each approach has limitations. However, without use of available technology to help prevent the marketing of modified foods with substantial allergenic potential, consideration for food labeling may be appropriate.

That is, labeling in the US is required 'if a food derived from a new plant variety differs from its traditional counterpart such that the common or usual name no longer applies, or if a safety or usage issue exists to which consumers must be alerted'. These issues are not lost on the general public, where allergenicity of modified foods is one concern among many worldwide dating back a number of years.

An examination of the amino acid sequence coded for by a transferred gene is recommended both in the situation where the gene transferred is from a known allergenic source and for instances where the gene transferred codes for a protein whose allergenic potential is largely unknown. This latter instance would be expected if the gene was obtained from an animal or plant source rarely consumed by the general population. The search for sequence similarity should include a review of similarity of allergens from both plant- and animal-derived foods, and inhalant allergens such as pollens, fungal spores, insect venoms and allergens which provoke immediate contact reactions. The chief difficulty in this approach is to define the number of contiguous amino acids required to create concern.

Because the optimal peptide length for binding appears to be between 8 and 12 amino acids for T cell epitopes and even longer for B cell epitopes, some have suggested that the sequence identity requires a match of at least 8 contiguous identical amino acids. There are certainly limitations in this approach. For example, not all food allergens have been sequenced, and this strategy will not identify confirmational or non-contiguous epitopes.

For a gene transferred from a known allergenic food or an allergenic food of concern, sequence similarity studies are strengthened by the determination as to whether a serum from individuals sensitive to the food of origin contains allergenspecific IgE which will bind with the gene product. These assays are usually performed as solid-phase immunoassays such as the RAST or RAST inhibition assay or the enzyme-linked immunosorbent assay (ELISA).

As a general procedure, this screening strategy for the transfer of a gene coding for an allergen gains strength when the number of sera from individuals sensitive to the food of origin is increased in number. Because in vivo skin prick tests have been held to be somewhat more sensitive than in vitro test results, some have suggested that if the in vitro tests are negative, skin testing should be performed.

Finally, double-blind placebo-controlled food challenges could be performed in controlled clinical conditions with patients sensitive to the food in question if the in vivo and in vitro tests are negative or equivocal. This latter consideration is one which must address a number of ethical issues, including the possibility of inducing anaphylaxis in test subjects, the availability of appropriate procedures, the availability of clinical safety data, utilization of institutional review board procedures, instructions to the subject to be tested, and what steps may be used to decrease risks to participants.

If the gene transferred is derived from a source where there is no general history of its use in the population and thus its allergenic potential cannot be assessed, the reactivity to specific IgEs from sensitive individuals cannot be generally applied. Thus in this situation, in addition to sequence similarity, stability to digestion may be employed and the issue of the use of an animal model to determine allergenicity becomes an issue. As has been discussed previously and as a general principal, food allergens appear to be resistant to acids and proteases. It has thus been argued that stability of food allergens to digestion may be used to help identify the products of transferred genes as to their possible allergenicity. However, insufficient information is available on possible differences in susceptibility to acid-denaturation and susceptibility to proteases between various allergenic food proteins and with proteins that possess weak or no allergenic potential to use this test in any other way but a relative indication of allergenic potential.

Any discussion of technical approaches to address the possible allergenicity of modified foods must consider animal models and their use in the determination of food protein allergenicity. Such an approach would help identify as to whether a gene product transferred from a protein where there is no history of exposure of that protein in the general population, is allergenic. Perhaps more importantly, a screening test to determine the allergenic potential of a modified food using an animal model would be ideal to address the question as to whether the genetic manipulation led to the expression or upregulation of an allergen in the recipient organism.

It is, of course, possible to make animals of various species allergic to specific proteins. The usual approach is to inject the protein in an adjuvant using the intraperitoneal route. For instance, the most commonly applied mouse model of asthma employs ovalbumin as its antigen, and the protein is administered intraperitoneally in alum. The difficulty arises in using sensitization by this means as a test for allergenicity, since it may be that both allergenic and non-allergenic proteins applied in this manner will provoke antigen-specific IgE, as determined by pulmonary or intradermal challenge, or using specific in vitro diagnostic assays for antigen-specific IgE. Indeed, it may be too much to expect any one animal model to predict allergenicity in humans of a specific food.

The differences between immune responses in an inbred population and in an outbred population are obvious. However, it may be reasonable to expect, with careful selection of the study animal, and the route and frequency of administration of the antigen, that an animal model might be developed that provides some ranking of allergenicity. That is, in such a model it would be expected that injection of an allergen such as Pen i 1 would provoke a stronger antigen-specific IgE response than would injection of a protein with low allergenic potential.

Reviews of the application of animal models to the identification of food proteins with allergenic potential look to the time when a specific animal model may prove of value in the evaluation of allergenicity. In the absence of a widely accepted animal model, it may be possible on a case by case basis to examine allergenicity of genetically modified foods, comparing the response in a specific animal model with the response that the animal exhibits to a non-allergenic food. In the genetic modification of foods directed at increasing crop production, modifying foods to increase their health benefits, decreasing a specific food's allergenicity, or increasing palatability, caution must be taken to avoid creating a new and significant allergenic food.

Those with food allergies mediated by IgE and immediate in onset should reasonably expect that application of technical approaches to monitor allergenicity will be employed in an attempt to identify and prevent the marketing of modified foods with significant allergenic potential. However, because almost any food may be allergenic in one or a very few individuals it is not reasonable to expect that modified foods will be absolutely and consistently without allergenic potential in everyone. It is reasonable to expect that the technical approaches available will help prevent the marketing of a modified food with significant allergenic potential.

Clearly, the modification of foods offers great promise in addressing worldwide issues such as increasing world grain production, adding specific nutrients to the diet, in producing edible vaccines and in reducing the allergenicity of existing allergenic foods. These beneficial aspects of technology applied to food modification should not necessarily be limited by concerns over allergenicity of modified foods, given the technical approaches available to monitor and help prevent the creation of new major allergenic foods.

## Chapter 10

# Regulatory for Novel Foods

### A EUROPEAN PERSPECTIVE

In Europe the sale of all novel foods, including genetically modified foods, is controlled by the EC Novel Foods Regulation. The Regulation establishes a European-wide pre-market approval system for novel foods and novel food ingredients, which is foods which have not been used for human consumption in Europe before, including those containing or produced from genetically modified organisms. It also establishes specific labelling rules which apply in addition to general food labelling requirements.

The main purpose of the legislation is to protect the consumer by requiring a rigorous safety assessment before a novel food can be approved for sale. A further important aspect of the regulatory framework is to ensure that consumers are able to make informed choices about foods that they eat. Although the Novel Foods Regulation came into effect relatively recently, some European member states, notably France, the Netherlands and the UK, had operated approval systems for many years. Indeed much of the EC novel food approval process is based on experience gained from operating the UK system.

The previous UK voluntary approval process The UK approval system for novel foods dates back to a system based on a voluntary arrangement with the food industry in 1980. Under the UK's previous voluntary arrangement for the safety assessment of novel foods, companies submitted applications for assessment by an independent advisory committee. Initially the Advisory Committee on Irradiation and Novel Foods was responsible for assessing applications, although in 1988 this committee was reconstituted into the Advisory Committee on Novel Foods and Processes (ACNFP).

To help companies identify the type of data that would be required to demonstrate that a novel food was safe, the committee developed a structured decision tree approach. The ACNFP updated this decision tree in 1994, using a series of linked questions which are designed to fully characterize the potential hazard of a novel food. The opinions expressed are those of the author, they are not official statements of the organization for which the author works. The main piece of European Community legislation that applies to novel foods is the EC Novel Foods and Novel Food Ingredients Regulations which came into effect on 15 May 1997. These regulations introduced a harmonized pre-market approval process for a wide range of novel foods, including genetically modified foods.

The regulations describe procedures for assessing the safety of novel foods and also contain specific labelling rules. Rules for the labelling of food containing ingredients derived from genetically modified soya and maize. Before describing the current Novel Foods Regulation in detail it is worth first considering the legislative framework that governs the release of genetically modified organisms into the environment. This framework is closely linked to the Novel Foods Regulation.

## Directive 90/220/EEC

In 1990 Europe adopted two directives which created a broad legislative framework covering the contained use, deliberate release and marketing of genetically modified organisms. Of this Directive 90/220/EEC controls the deliberate release into the environment of genetically modified organisms.

The directive is intended to protect human health and the environment and to establish a single market for products containing genetically modified organisms. It defines a genetically modified organism as 'an organism in which the genetic material has been altered in a way that does not occur naturally by mating and/or natural recombination' and includes a non-exhaustive list of techniques that are regarded as genetic modification. The directive covers both the release for experimental purposes and the commercial sale of genetically modified organisms.

However, Article 10 of the directive makes provision for product-specific regulations requiring a similar environmental risk assessment, such as the Novel Foods Regulation, to replace the need for a marketing consent under Directive 90/220/EEC. Under the directive, all releases to the environment, whether for experimental purposes or for marketing, must be approved by the competent authority in advance.

An applicant is required to provide all the necessary information set out in the directive, including information on the nature and stability of the genetically modified organism itself, on the potential receiving environment and on the interactions between the genetically modified organism and the environment. The authorities have the power to request further information as required. In addition an applicant is required to provide a statement assessing the risks that the genetically modified organism poses to human health and the environment. This risk assessment is then evaluated by the competent authority to which the application is made. In the case of experimental releases, a consent is issued by the member state where the trial is to take place.

A summary of the information provided, together with the decision, is circulated to other states of the European Community for information only. They cannot intervene in the decision but they may comment. This allows competent authorities to benefit from the experience of others, and to indicate any concerns if there is the likelihood of cross-border effects.

In order to verify the assumptions of the risk assessment and to enable informed decisions to be made on future applications, a consent holder is required to provide a report of the release and details of any relevant information obtained from the release that relates to risks to human health or the environment. Similar procedures apply in the case of a request to market a genetically modified organism. However, the consent, when issued, applies across the whole of the European Community. Accordingly once an application has been considered by the competent authority of the member state in which the product is first to be marketed, all other member states have the opportunity to comment or raise objections.

If the initial assessment is favourable and no objections are raised the initial competent authority formally issues the marketing consent. If objections are raised, recent practice has been that they are referred by the European Commission to one of their scientific committees before the Commission submits a proposed decision to the 'Article 21' regulatory committee where decisions are taken by qualified majority voting. If the proposal is not adopted it is referred to Council where unanimity is required to amend the European Commission proposal. In March 2001 Directive 2001/18/EC was adopted.

Amongst the provisions in the new directive consents to market genetically modified organisms will be limited to a maximum of 10 years, during which time the notifier will be required to comply with a monitoring plan described in the consent. The purpose of the monitoring plan is to confirm that assumptions regarding the occurrence and impact of potential adverse effects of the genetically modified organism or its use made in the risk assessment are correct and to identify the occurrence of adverse effects of the genetically modified organism or its use on human health or the environment which were not anticipated in the original risk assessment.

## WHAT IS A NOVEL FOOD

The scope of the Novel Foods Regulation is defined in Articles 1 and 2 of the Regulation. In Europe a novel food is defined as a food or food ingredient which has not hitherto been used for human consumption to a significant degree within the Community and which falls within one of the following categories.

- Foods and food ingredients containing or consisting of genetically modified organisms within the meaning of Directive 90/220/EEC;
- Foods and food ingredients produced from, but not containing, genetically modified organisms;
- Foods and food ingredients with a new or intentionally modified primary molecular structure;
- Foods and food ingredients consisting of or isolated from microorganisms, fungi or algae;
- Foods and food ingredients consisting of or isolated from plants and food ingredients isolated from animals, except for foods and food ingredients obtained by traditional propagating or breeding practices and having a history of safe food use;
- Foods and food ingredients to which has been applied a production process not currently used, where that process gives rise to significant changes in the composition or structure of the foods or food ingredients which affect their nutritional value, metabolism or level of undesirable substances.

Accordingly products which had been approved under previous national procedures, but which had not been marketed are subject to the provisions of the regulation. Food additives, flavourings used in foodstuffs and extraction solvents used in the production of foodstuffs are excluded from these regulations on the basis that such products are already covered by existing community legislation.

Indeed the Novel Foods Regulation specifically states that such exemptions only apply as long as such products are the subject of a comparable safety assessment. The Novel Foods Regulation also makes provision for decisions to be taken on a case-by-case basis as to whether a specific food falls within the scope of the regulation. Member states have received many requests from potential applicants for clarification of the status of specific products. Most of these requests relate to dietary supplements, of which many are freely available, where under the Dietary Supplements and Health Education Act 1994, they are exempt from any pre-market approval process. In Europe, if they do not have a significant history of consumption they require approval before they can be placed on the market. Although the European Community definition of a novel food is somewhat general, there are several examples that help to clarify its scope.

In April 1996 Monsanto's glyphosate-tolerant soya was approved for importation and processing to non-viable products following a rigorous safety assessment under Directives. Similarly, in February 1997 approval was issued for importation and cultivation of Ciba-Geigy's insect-resistant and glufosinate ammonium-tolerant maize. Both of these products, which were first grown commercially in the USA in 1996, had been consumed in significant quantities (greater than 1 million tonnes) by May.

Accordingly, member states ruled that they were outside the scope of the Novel Foods Regulation, and in any case they had both been fully assessed for food safety. Although, as will be discussed later, it was agreed that for labelling purposes they should be treated in the same way as products approved under the Novel Foods Regulation. Other products that have been judged to fall outside the scope of the Regulation include the mycoprotein Quorn, first approved in the UK in 1983 and sold in most member states since then. Lactulose was also considered not to be novel on account of its use in biscuits in one member state.

A more recent example which caused considerable debate was a margarine-based product containing plant sterols. This product, which had been sold in significant quantities in Finland since its introduction in 1995, was intended to reduce cholesterol absorption. Again member states ruled that the product was not a novel food. However, a similar product developed by another company was considered for approval under the Novel Foods Regulation because it had not been on sale before May 1997.

It is possible to draw some conclusions as to what comes within the scope of the Regulation. Any product sold to consumers in one or more member states prior to May 1997 is not a novel food. If the food or food ingredient has not been sold in Europe, and is not a food additive, flavouring substance or extraction solvent, it is subject to the provisions of the Regulation.

The foods generating most enquiries as to their status under the Novel Foods Regulation are dietary supplements. It is clear that the regulation will have a significant impact on this sector with companies now being required to generate the necessary safety data to support an application for approval under the Regulation.

## Approval Procedures for Novel Foods

Article 3 of the Regulation summarizes the procedures that apply to all foods covered by the Regulation. Before any novel food can be sold in Europe, member states must be satisfied on the basis of data provided in an application that the food does not:

- Present a danger to the consumer
- Mislead the consumer,
- Differ from foods or food ingredients which they are intended to replace to such an extent that their normal consumption would be nutritionally disadvantageous for the consumer.

Articles 4-9 of the Novel Foods Regulation describe the procedures that are followed in considering all applications, or in the case of Article 5, simplified procedure notifications. Each member state is required to establish a competent authority for the receipt and processing of applications. In the UK the competent authority is the Food Standards Agency.

In addition, each member state is required to appoint a food assessment body to evaluate applications. In the UK this task is done by the independent Advisory Committee on Novel Foods and Processes, although this committee can seek advice from other expert committees such as the Food Advisory Committee, the Committee on Toxicity or the Advisory Committee on Releases to the Environment. The publication of guidelines which describe the information necessary to support an application, the presentation of such information and the format for assessment reports prepared by member states.

These guidelines will be described later. A company wishing to obtain approval for a novel food in Europe is required to submit two copies of their application; one copy goes to the member state where the product is intended to be marketed for the first time (the initial member state), the other copy is sent to the European Commission. The application consists of a summary document, copies of all study reports and any other supporting data and a proposal for the labelling of the product.

Upon receipt of an application the initial member state first checks that the application is accompanied by all the necessary supporting data as identified by the guidelines. Once satisfied that an application is complete, the initial member state informs the European Commission and other member states that an application has been accepted. The date of the notification to the European Commission is taken as day 1 of the 90-day assessment period. At the same time the applicant's summary document is circulated to all member states. It is in the applicant's interests to ensure that their summary document lists all the supporting data and is sufficiently detailed to enable other member states to form an opinion in the absence of the supporting data.

However, it is recognized that member states will have the option of requesting some or all of the supporting data if necessary. Within 90 calendar days the initial member state is required to forward a copy of its assessment to the European Commission, The European Commission then forwards the initial assessment report without delay to all member states for their consideration. If during the initial assessment it becomes necessary to obtain more information, the applicant will be asked to provide the necessary additional information; during this period the 90-day clock is stopped.

The initial member state is required to inform the Commission and other member states of the reason for the delay immediately. Similarly the European Commission and member states shall be informed when the clock is restarted. While recognizing that stopping the clock is necessary where further information is required, member states endeavour, as far as is possible with a quick scrutiny, to ensure that an application is complete before accepting it so as to keep the need for clock stopping to a minimum.

From the date the European Commission circulates the opinion of the initial member state the remaining member states have 60 calendar days in which to submit any comments or reasoned objections to the European Commission. There is no provision for stopping the clock within the 60-day period. Where member states raise objections they may seek to resolve them on a bilateral basis with the initial member state or the applicant.

If at the end of the 60-day period for comment the initial member state has received no reasoned objections, and no objections have been received by the European Commission, the applicant is immediately informed in writing by the initial member state that they may place their product on the market. If reasoned objections are received, which cannot be resolved on a bilateral basis, the initial member state shall write to the applicant informing them that an 'authorization decision' in accordance with Article 7 of the Regulation is required.

If an authorization decision is required, a decision on whether or not to issue an approval is taken centrally. The first step of the authorization procedure involves a meeting of the novel foods competent authorities where objections raised by member states are discussed. At this point member states decide that either the application requires a more detailed safety assessment, in which case th application is referred to the Scientific Committee for Food i accordance with Article 11 of the Regulation, or that the applicatio does not raise any issues that have an effect on public health, in whic case a decision as to whether the product should be allowed onto th market should be put to a vote at the next meeting of the Standin Committee for Foodstuffs, the body described in Article 13 of th Regulation. If the novel foods competent authorities meeting decid that the Scientific Committee for Food needs to consider the safety o the product in more detail, they may wish to consider which aspect of the application the Scientific Committee for Food should be aske to address. Once the advice of the Scientific Committee for Food i available, the European Commission is required to put a proposal fo an authorization decision to the Standing Committee for Foodstuff for a vote. The proposal will specify the conditions of use of the foo or food ingredient; the designation of the food or food ingredient, an its specification; and specific labelling requirements for the food. If th Standing Committee for Foodstuffs decides by a qualified majority t issue an authorization decision, the European Commission informs th applicant in writing immediately.

A copy of the decision is also published in the Official Journal o the European Communities. If no agreement is reached in the Standin Committee for Foodstuffs, the matter is referred to the Council o Ministers. The Council has three months in which to act, eithe adopting the European Commission proposal by qualified majorit voting or amending it by unanimity. If it fails to act within the thre months the European Commission proposal is adopted by default.

30 applications had been received, of which six had completed al stages of the approval process resulting in the approval of four product and the rejection of two products on the basis of incomplete supportin data. So far there have been 11 applications for foods derived fron genetically modified crops, including processed tomato products chicory, maize and soybeans. Other applications have included plan and dried leaves from Stevia rebaudiana, phospholipids from egg yolk yellow fat spreads with added phytosterol-esters and the use of higl pressure to treat fruit preparations.

## Approval of foods Containing or Consisting of Genetically Modified Organisms

As discussed earlier, Directive 90/220/EEC provides a Europea regulatory framework to protect human health and the environmen from the introduction of genetically modified organisms. Article 9 o ıe Novel Foods Regulation removes the requirement to obtain a ıarketing consent, under Directive for foods which contain or consist f genetically modified organisms. In support of an application for such food, a company is required to submit copies of any consents already btained under Directive 90/220/EEC for research and development urposes.

In addition an applicant must provide all the relevant information ıat would have been required under that directive to ensure that all ecessary measures are taken to safeguard both human health and the ıvironment. An application under Directive must be made for those ıses in which a genetically modified organism is intended for uses ːher than food, such as animal feed or cultivation within Europe, in ːder to secure scrutiny of human health and environmental effects.

## nk with the EC Common Catalogue for Seeds

Before crops can be grown for commercial purposes in Europe they ave to be included on the National List of a member state or the EC ommon Catalogue. Various European Community seeds directives ːquire that varieties must be grown in official tests and trials, normally r two years, to establish that they are distinct, uniform and stable ıd have value for cultivation and use. Any seed marketed must be ːrtified as to its varietal purity and germination standard.

These requirements apply to any new variety, whether produced / conventional means or the use of genetic technology. In order to ısure that, once a final approval for food use of a genetically modified op is obtained the seed can be offered for sale to farmers without rther delay, companies typically enter a variety into National List ials under an experimental trials consent issued under Directive 90/ 20/EEC. Addition to the EC Common Catalogue, which normally llows National Listing, allows the variety to be grown commercially ıywhere in the European Community. The intention of the link to e seeds directives in the Novel Foods Regulation was to enable a ımpany to submit a single application for approval to grow a ːnetically modified crop commercially in Europe in accordance with irective 90/220/EEC; to sell the food in accordance with the Novel ıods Regulation and to have the variety entered onto the EC Common atalogue. In practice this linkage has never been used and is likely to removed when the Novel Food Regulation is revised.

## mplified Procedures

Articles 3 and 5 of the Regulation contain provisions for companies notify the European Commission when they first place a product on the market which is considered to be 'substantially equivalent to existing foods or food ingredients in terms of composition, nutritional value, metabolism, intended use and level of undesirable substances'. Under the regulations companies have the option of submitting a notification based on the opinion of one competent authority that the particular food or food ingredient is 'substantially equivalent' to existing foods or food ingredients.

Alternatively a company can submit a notification based on readily available and generally recognized scientific evidence such as papers from peer reviewed journals. Some companies have notified the European Commission that they have placed on the market various processed products derived from varieties of genetically modified maize and refined oil from herbicide-tolerant oilseed rape. The European Commission publishes a list of notifications in the C series of the Official Journal of the European Communities each year. A total of 11 notifications had been submitted under the simplified procedures by June 2001. However, particularly in the case of oilseed rape it is not clear whether the oil has in fact been marketed, given that at least one variety of oilseed rape has yet to be grown commercially.

With the exception of oil from three varieties of oilseed rape, all the notifications submitted to date have been supported by a full safety assessment done prior to the novel foods regulation coming into force. The simplified procedure has caused considerable confusion by use of the term 'substantially equivalent'. This has frequently been confused with the concept of substantial equivalence, which is used to structure the safety assessment process. Although member states recognize that the simplified procedure is an option available to companies, all member states agreed with the UK position that only highly processed, refined foods derived from genetically modified crops, such as hot pressed oil, white sugar and starch hydrolysates, would be suitable for consideration under the simplified procedures on the grounds that neither DNA nor protein resulting from the modification would be expected to be present. In submitting a notification under the simplified procedure a company has to demonstrate that no genetically modified material (protein or DNA) is present in the food ingredient. Nevertheless, this simplified procedure has not been well received in many countries and is likely to be phased out when the novel food regulations are revised.

## Labelling

In reaching agreement on the final text of the Novel Foods Regulation, it was the issue of labelling which proved most difficult to reconcile from all viewpoints. A significant contribution to the debate at the time was made by the Group of Advisers on the Ethical Implications of Biotechnology of the European Commission when in May 1995 they published an opinion on 'Ethical aspects of the labelling of foods derived from modern biotechnology'.

Amongst the recommendations that this group made was that consumers must be provided with information which, for transparency should be:

- Useful, adequate and informative;
- Clear, understandable, non-technical;
- Honest, not misleading or confusing, and which aims to prevent fraud;
- Enforceable, i.e. possible to verify.

They also recommended that consumers have a right to be able to make informed choices about what they eat; so they can legitimately expect to receive a clear indication of where additional information can be obtained, especially when their choices include cultural and religious considerations. Article 8 of the Novel Foods Regulation describes the specific labelling requirements that are required for novel foods in addition to the existing European Community law on the labelling of foodstuffs, the main source of which is contained in Directive. The main provisions of this directive, which applies to all foodstuffs placed on the market in the European Community, are to ensure that labelling does not mislead the consumer to a material degree as to the characteristics of the foodstuff, or by attributing properties to it which it does not have, or by suggesting that it has special characteristics when in fact all similar foods possess such characteristics.

This directive applies to foods ready for delivery to the ultimate consumer or to mass caterers. The labelling requirements contained in the Novel Foods Regulation apply to food sold to the final consumer.

*They cover the following groups of foods*:

- Foods and food ingredients obtained from genetically modified organisms when, on the basis of a scientific assessment, they are judged not to be equivalent to an existing food.
- Novel foods that contain material which is not present in an existing equivalent foodstuff and which may have implications for the health of some sections of the population. An example would be a protein from a known food allergen source such as peanuts.
- Novel foods that contain material which is not present in an

existing equivalent foodstuff and which gives rise to ethical concerns.

- All foods that contain or consist of genetically modified organisms within the meaning of Directive.

In addition, if the novel food has no existing equivalent the Regulation contains powers for the adoption of provisions to ensure that consumers are adequately informed of the nature of the food ingredient concerned. An example of where such a provision might be required is in the context of low-calorie fat replacers, although under Directive 90/496/EEC on nutrition labelling the energy value for all fats is set at 9kilo calories/gram (37 kJ/g) for the purpose of calculating energy values for the final foodstuff.

Although the Regulation makes provision for labelling where a food contains material which might give rise to health implications for groups of the population, it presupposes that such a product would be approved for sale. In the case of a food containing a major allergenic protein, it is unlikely that such a product would be approved. In 1992 the then UK Minister of Agriculture, Fisheries and Food appointed a Committee under the chairmanship of the Reverend Dr J Polkinghorne 'to consider future trends in the production of transgenic organisms; to consider moral and ethical concerns (other than those related to food safety) that may arise from the use of such organisms and to make recommendations'.

In its report, one of the Committee's recommendations was that food should be labelled if it contains copies of ethically sensitive genes (human genes or genes of religious significance) or if a plant contains copies of an animal gene. This recommendation, and a similar recommendation from the European Commission's group of ethical advisers is reflected in the requirement to label if a food contains material which is not present in an existing equivalent foodstuff and which gives rise to ethical concerns.

In Europe, since the Novel Foods Regulation came into force the debate over the extent to which foods derived from genetically modified organisms should be labelled has been intense. The current labelling rules, which were agreed by all member states, reflect the view that where measurable differences in composition, compared with a non-modified counterpart exist, labelling should be required to enable consumers to make an informed choice.

In this respect the labelling rules contained in the Novel Foods Regulation itself are not absolutely precise. However, the situation was clarified with the introduction of detailed rules for the labelling of genetically modified ingredients derived from Monsanto's soya and

Ciba Geigy's maize contained in EC Regulation. These rules require the labelling of foods sold to the final consumer containing ingredients derived from Monsanto's soya or Ciba Geigy's maize, in which either DNA or protein arising from the genetic modification is present.

This approach was seen as setting a precedent for the labelling of all genetically modified foods approved under the Novel Foods Regulation. Nevertheless, since the introduction of Regulation 1139/98 no GM foods have been approved for sale in Europe. During negotiations on Regulation it was accepted that, although tests for DNA using the polymerase chain reaction were more sensitive than tests for proteins, labelling should be triggered by the presence of either protein or DNA resulting from a genetic modification.

If a company or enforcement authority obtains a negative result with one method they would need to follow this up with the other method. However, if the first method used gave a positive result there would be no need for re-testing. In the preamble to the Novel Foods Regulation itself there is a clause which would in effect enable companies selling bulk consignments of commodity crops to label the crop 'May contain genetically modified material'.

During negotiations on the soya and maize labelling rules member states were determined to remove this option, which was seen as providing no meaningful information to the consumer. Accordingly, Regulation 1139/98 requires a definitive statement 'produced from genetically modified soya' or 'produced from genetically modified maize' whenever DNA or protein resulting from the modification is present. Because products containing soya and maize were already on sale in Europe before labelling rules were agreed it was recognized that it would be impractical to require products already on sale to be re-labelled. Consequently the regulation only applies to foods manufactured and labelled after 1 September 1998.

Furthermore, since European Community labelling policy is focused at the level of the ingredient rather than the whole food product, it was intended that enforcement would also be at the level of ingredients used in a foods production, thereby avoiding the need to develop detection methods capable of detecting protein or DNA in a finished food product. In the preamble to Regulation 1139/98 there is a recognition that the labelling requirements should be no more burdensome than necessary but sufficiently detailed to supply consumers with the information they require.

In order to avoid the need for enforcement authorities to sample those ingredients which, by virtue of the degree of refining during production, will not contain protein or DNA, the regulation makes provision for the establishment of a so-called 'negative list'. This was intended to be a list of those ingredients derived from soya and maize which have been clearly demonstrated to contain neither protein nor DNA; however, the list has never been developed.

Recognizing that there is a demand for supplies of non-genetically modified ingredients as a means of providing consumer choice, a number of European retailers have started to obtain supplies of soya and maize from identity preserved sources. The UK Government encouraged the provision of alternative supplies in 1998 when the then UK Ministry of Agriculture, Fisheries and Food published a list of suppliers offering non-genetically modified material to assist those manufacturers and retailers who wished to offer their customers a choice.

Companies seeking to obtain supplies from identity preserved sources found that no company would offer an absolute guarantee that no genetically modified material had become accidentally mixed with the consignment during the supply chain. Regulation 1139/98 recognized this difficulty and proposed the development of a low-level de minimis threshold below which labelling would not be required. In April 2000 EC Regulation came into effect.

This Regulation amended EC Regulation 1139/98 by extending the requirements to foods sold to mass caterers and setting a de minimis threshold of 1% for the adventitious contamination of non-GM material. For such ingredients there is no need to label them as GM if they contain less than 1% GM material. The threshold applies only to ingredients obtained from non-GM sources; this flexibility does not apply to supplies obtained from sources of unknown origin.

Companies also need to demonstrate that their ingredients are of non-GM origin, and it is possible that the use of documented and audited identity preservation systems could satisfy this requirement. Steps should also be taken to keep the level of adventitious contamination in non-GM supplies to a minimum. Although the level agreed for the threshold is 1%, in practice the need to provide proof that ingredients are of non-GM origin should ensure that actual levels are kept well.

It is also important to realise that because the limit operates at the level of each individual ingredient, the actual amount in the final foodstuff will be much lower. A consequence of the fact that Regulation 1139/98 was made under Article 4 of the food labelling Directive 79/112/EEC is that member states have the flexibility to exempt or otherwise from labelling provisions made under the directive those suppliers offering non-prepacked foods to the final consumer.

In keeping with its policy of ensuring that all foods containing genetically modified material are clearly labelled, the UK Government introduced national regulations in 1999 requiring all businesses selling non-prepacked foods to the final consumer to ensure that information as to the presence of genetically modified soya or maize is provided where consumers request it.

In April EC Regulation also came into effect, requiring the labelling of foods containing additives and flavourings that had been obtained from genetically modified organisms where novel protein or DNA was present in the final food. There is currently no requirement to label additives and flavouring substances obtained from genetically modified organisms sold as such to the final consumer (a very minor use), since this would require amendment of the additives and flavouring framework directives.

In December the European Commission announced that it was working on proposals for a regulation requiring the traceability of genetically modified organisms and food and feed ingredients derived from them. A second proposal, for a regulation for the approval of genetically modified food and animal feed, is expected to contain a requirement that all ingredients derived from genetically modified organisms are labelled, regardless of whether any genetically modified material can be detected in the final food. In taking these proposals forward, careful consideration will need to be given to their practicality, proportionality and enforceability.

## Guidelines Accompanying the Novel Foods Regulation

In order to identify the type of information that potential applicants would be required to submit in support of an application and to ensure a consistent approach to the assessment of an application by all member states, the European Commission published guidelines, developed by their Scientific Committee for Food, as a European Commission Recommendation.

These guidelines draw heavily on the structured approach to the safety assessment of novel foods developed by the ACNFP and first published in 1991 and revised in 1994. The ACNFP approach and the subsequent Scientific Committee for Food's guidelines use a series of linked questions to ensure that potential hazards of a novel food are fully characterized. Underlying the structured questions there are a number of key issues which need to be addressed in assessing the safety of a novel food. First and foremost is the recognition that foods are frequently a complex mixture of many thousands of chemical entities. Furthermore, foods are generally ingested in much higher amounts than would be expected for discrete chemical entities such as food additives or pharmaceuticals. For compounds such as pesticides, pharmaceuticals, industrial chemicals and food additives, animal studies represent a major element in the safety assessment. In such cases, the test substance is generally well characterized, of known purity, of no nutritional value and human exposure is generally low.

It is therefore relatively straightforward to feed such compounds to animals at a range of doses, some several orders of magnitude greater than the expected human exposure levels, in order to identify any potential adverse effects of importance to humans. In this way it is possible, in most cases, to determine levels of exposure at which adverse effects are not present, and so set safe upper limits by the application of appropriate safety factors.

By contrast, foods are complex mixtures of compounds characterized by wide variation in composition and nutritional value. Due to their bulk and effect on satiety they can usually only be fed to animals at low multiples of the amounts that might be present in the human diet. In addition, a key factor to consider in conducting animal studies on foods is the nutritional value and balance of the diets used, to try to avoid the induction of adverse effects which are not related directly to the material itself.

Picking up any potential adverse effects and relating these conclusively to an individual characteristic of the food can therefore be extremely difficult. It was very much in recognition of these difficulties that the concept of substantial equivalence was developed as a framework for the safety assessment of genetically modified foods in particular. The guidelines also attach considerable importance to compositional analysis, particularly in respect of key nutrients and natural toxicants.

However, the guidelines provide no real insight into how such compositional data should be generated to ensure that safety assessments are based on statistically valid data. The UK has issued a paper to prompt further discussion of the subject. Given the importance attached to the comparative approach to safety assessment of foods derived from genetically modified crops it is important to ensure that any real differences are not masked by natural variation.

This natural variation may be large because of the wide variation in conditions, such as soil fertility, temperature, light or water availability, under which crops are grown. The issue may be compounded by other effects including agronomic treatments to the growing crop, such as herbicide applications to a herbicide-tolerant crop, or post-harvest stress during storage (e.g. for potatoes stored in clamps). To minimize the possibility that any changes caused by a crop modification might be masked by natural variation in a crop, the UK has made a number of recommendations on the design of field trials. For crop varieties which are intended to be grown commercially in Europe, the location of trial sites should be representative of the range of environmental conditions under which the varieties would be expected to be grown.

The number of trial sites should be sufficient to allow accurate assessment of agronomic and compositional characteristics over this range. Similarly, trials should be conducted over a sufficient number of years to allow adequate exposure to conditions met in nature. The UK has recommended that the minimum number of trial sites should be six and the minimum number of years should be two.

To reduce any effect from naturally occurring genotypic variation within a crop variety, and to minimize the impact of environmental effects at a given trial site, the minimum number of replicates used should be not less than three and will need to be appropriate to the crop species under trial. To allow effective comparison the UK also recommends that a genetically modified crop and its non-modified counterpart should be grown in adjacent plots at the same site and at the same time.

Information on likely intakes of a novel food is also of special importance in assessing its safety. In many instances the introduction of a novel food will not alter overall intakes of a particular dietary component. For example, the introduction of oil from genetically modified herbicide-tolerant oilseed rape is unlikely to alter the overall intake of oilseed rape oil. However, the introduction of a foodstuff fortified with plant sterols and sold on the basis of claims that it can help reduce cholesterol levels may result in a significant increase in the consumption of plant sterols.

Where the introduction of a novel food is expected to give rise to significant shifts in consumption patterns the novel food guidelines recommend the introduction of a surveillance programme to accompany a products launch. In reaching conclusions on the overall safety of a novel food it is important to recognize that the assessment takes into account both food safety and nutritional safety implications. The latter can be of particular importance when considering the implications for groups such as infants, children, pregnant and lactating women, the elderly and those with chronic diseases that have an impact on nutritional status. Two other aspects of the safety assessment which are considered elsewhere relate to the assessment of the allergic potential of a novel food and, in the case of genetically modified foods, the safety implications of any selective marker genes remaining in a crop as a result of the modification. In designing a series of structured questions to cover all categories of novel foods whilst keeping the decision tree as simple as possible, the guidelines identify six classes of novel foods which present similar safety assessment issues.

*The six classes are*:

- Pure chemicals or simple mixtures from non-genetically modified sources. This class identifies data requirements for some foods falling within the third, fourth and fifth categories of the regulation and would include products such as low-calorie fat replacers.
- Complex novel foods from non-genetically modified sources. This class identifies data requirements for foods derived from whole plants, animals or micoorganisms which fall within categories 4 and 5 of the Regulation. This class would include products such as stevia leaves.
- Genetically modified plants and their products.
- Genetically modified animals and their products.
- Genetically modified microorganisms and their products. Classes 3-5 cover both viable genetically modified organisms and foods derived from them. These classes would include genetically modified chicory and food products derived from genetically modified processing tomatoes.
- Foods produced using a novel process. This category covers foods covered by category 6 of the Novel Foods Regulation and would include processes such as high-pressure processing or new processes catalysed by enzymes.

*The first five classes are then subdivided into*:

- Those foods for which there is a history of consumption of the source of the food, or in the case of genetically modified foods the host organism;
- Those foods for which there is no history of food use of the novel food or the host organism.

Having assigned a novel food to one of six classes, the guidelines identify the following information requirements which apply to some, or all classes of novel foods:

- Specification of novel food
- Effect of production process applied to novel food

**History of Source Organism**

- Effect of genetic modification on properties of host organism
- Genetic stability of the genetically modified organism

- Specificity of expression of novel gene
- Transfer of genetic material from genetically modified microorganisms
- Ability to survive in and colonize gut
- Anticipated intake/extent of use
- Information on previous human exposure
- Nutritional information
- Microbiological information
- Toxicological information.

Each information requirement consists of a series of structured questions that an applicant needs to address in order to provide all the necessary information to support their responses as part of the application package. In submitting an application companies are advised to follow the recommended format contained in part II of the guidelines.

In addition to providing details of the product, the applicant is advised to provide sufficient data to enable a food to be assigned to one of the six categories of novel foods set out in Article of the Regulation. In working through the decision trees in the guidelines an applicant should provide all relevant data used to support the response to each question.

All labouratory data should be generated in accordance with the principles of good labouratory practice. The application should consist of a summary document, full study reports for all supporting data and a proposal for the labelling of the food.

## Transparency

Article 10 of the Novel Foods Regulation makes provision for detailed rules on data confidentiality to be adopted. To date, no such rules have been adopted and member states have introduced their own procedures. In the UK the ACNFP has continually sought to increase the transparency of its decision making.

Since 1988 the ACNFP has published annual reports which have included copies of all the assessment reports prepared during that year. More recently the ACNFP has published agendas and full minutes of their meetings as well as annual reports.

## Post-market Monitoring

Article 14 of the Novel Foods Regulation requires the European Commission to monitor the application of the Regulation and its impact on health, consumer protection, consumer information and the functioning of the internal market. Whilst recognizing that novel foods are only approved for sale within the European Community following a rigorous safety assessment following the procedures the UK has been considering proposals for the introduction of a post-market monitoring system for novel foods.

Such a system would be very much an addition to the existing safety assessment procedures rather than an alternative. For post-market monitoring to be practical it is necessary to obtain data on consumption patterns and to relate these to any reported adverse health effects.

The ACNFP has been looking at ways of integrating the two components necessary for such a system to operate. In terms of mechanisms for identifying patterns in health effects, the UK has an extensive array of health databases which enable disease patterns to be tracked over time and analysed for any regional differences. The other vital component is an ability to obtain details of what foods people have eaten and whether there have been any shifts in their consumption patterns. A number of sources already exist for food consumption data within the UK.

Article of the Novel Foods Regulation already contains a provision for action to be taken if new information suggests that a food approved under the Regulation might be endangering human health or the environment. If such a situation were to occur a member state may temporarily restrict trade in the food while the matter is considered by the Standing Committee for Foodstuffs.

## Enforcement

Although both the Novel Foods Regulation and the regulation on the labelling of genetically modified soya and maize are directly applicable and legally binding in all member states, each member state is responsible for enforcement in its territory.

Under the directive on the official control of foodstuffs member states are required to submit an annual return to the European Commission giving details of the number and type of inspections carried out and details of any infringements found. Although the European Community Novel Foods Regulation only came into effect on 15 May, it builds on many years' experience within some member states.

In addition, in the case of genetically modified foods the regulation builds on experience gained from the approval for sale of genetically modified organisms under Directive. In addition to introducing a uniform approach to ensure that novel foods are rigorously assessed for safety to ensure consumer protection, the Regulation also makes special provision for labelling requirements to be included, where appropriate, as a condition of approval. This is particularly important in the case of genetically modified foods to ensure that consumers are able to make informed choices about foods that they eat. The Novel Foods Regulation has now been in force for four years, and it is clear that it is not operating effectively. Although initial problems with the operation of the Regulation were attributed to a lack of collective experience between member states, it soon became apparent that the problems were deeper.

## Chapter 11

# Analysis of Plant Genome

Molecular makers are useful tools for developing detailed linkage maps of species that previously were very poorly mapped. These maps have obvious utility for the identification of markers linked to genes of agronomic or horticultural importance.

*Beyond this practical utility, molecular markers and molecular maps can provide detailed information regarding*:

- The structure of the genome of a species
- The evolutionary relatedness of species within a family
- sequence rearrangements during evolution
- The relationship between physical and genetic distances linkage drag

## PLANT GENOME STRUCTURE

The genomes of all eukaryotic species consist of single-copy, middle repetitive and high copy number sequences. To gain an understanding of the per centages of each of the classes within any particular species, a group of random clones can be hybridized to blots of plant DNA. One such study was performed by Zamir and Tanksley. They hybridized 50 random genomic clones to tomato DNA. Washing was performed at two stringencies. The following are the results with regards to copy number.

| Clone Class | Low Stringency Wash | High Stringency Wash |
|---|---|---|
| Single copy clones | 44% | 78% |
| Multiple copy clones | 46% | 18% |
| Repetitive clones | 10% | 4% |

The table shows that hybridization stringency has a significant effect upon the number of sequences to which a sequence hybridizes. At the higher stringencies, most of the clone recognized only a a single copy within the genome. What this also shows is that the tomato genome contains many sequences that are about 80% homologous, but fewer sequences that are highly homologous. Clearly similar sequences must have diverged by some mechanism during the evolution of the species. The authors also hybridized these 50 clones to filters containing tomato, related tomato species, eight Solanum species and one member of the Curcurbitaceae family.

As a control,single copy tomato cDNA clones were also hybridized to these clones. At moderate stringency, the cDNAs hybridized to all the tomato species, and most hybridized to the Solanum species. 80% of the clones hybridized to the related Curcurbitaceae species. In contrast, only 50% of the random clones hybridized to related tomato and Solanum species, and only 10 % hybridized to the Curcurbitaceae species.

The principal conclusion that can be drawn from these hybridizations is that the random sequences (as represented by the random genomic clones) are evolving faster than the single-copy sequences. Why? At the higher stringency, homology to distant related species was reduced. Therefore these sequences have undergone greater divergence. This evidenced by the fact that 0/5 of the repetitive sequences hybridized to tobacco, whereas 20/45 of the single and multiple copy clones hybridized to tobacco.

## EVOLUTIONARY RELATEDNESS

The typical experiment begins with a large collection of cultivars or species. These samples are then analysed by hybridization with RFLP clones or more recently by the PCR-RAPD amplification. The similarity of each pair of samples is measured by calculating the number of common bands or amplification products. One estimator was developed by Nei and L. The formula is:

$$F = 2nXY/(nX + nY)$$

where nX and nY are the total number of fragments for sample X and Y and nXY is the number of fragments shared by the samples. Data is normally collected for a relatively large number of hybridizations or PCR-RAPD amplifications. The similarity data is then used in a cluster analysis to develop dendrograms which show the molecular relatedness of the species.

*These types of analyses can*:

- Provide independent support to previous phylogenetic and evolution hypotheses.
- Identify gene pools within a genus.
- Estimate genetic diversity within a genus.
- Provide necessary data to select appropriate parents for a molecular mapping project.

## COMPARATIVE GENOME MAPPING

Often the clones used to identify RFLPs in one species can be used in a second species. These heterologous probes can then be ued to develop a RFLP map in the second species. Once the two species have been mapped, the relative evolution of the two species can be compared. Comparative mapping can identify inversions, translocation and duplications that have occurred. Genetic factors can also be assessed by comparing map distances of genes with conserved gene order in the two species.

A comparison of the related grass species sorghum and maize was made using maize clones. The diploid chromosome number of the two species is twelve. Many of the sorghum chromosomes contained regions from two of the maize chromosomes. This result may represent the ancestral duplication of chromosomal material. Furthermore, during maize evolution duplicate genes have also occurred to a greater extent than that seen in sorghum. Only nine inversions of gene order have been observed between the species. Maize and sorghum were next compared with regard to genetic distance. Conserved gene orders were compared, and these linkages measured 862 cM in maize and 835 cM in sorghum. Therefore, it can be concluded that since the divergence of sorghum and maize, large chromosomal changes have not occurred and much of the recombination distance has been maintained.

The same type of analysis was performed with two more distant species, tomato and pepper. Little linkage conservation appears to have been maintained since the divergence of these two species, even though the chromosome number has been conserved. The largest conserved linkage block is a 63 cM block of tomato chromosome two that was located on chromosome I of pepper. Some chromosomes of pepper contained six distinct regions of the tomato genome (pepper chromosome X). These chromosomal breakages and rearrangements are not centromeric in nature, suggesting that evolution involved breakage throughout the genome. Because duplicated regions were found in pepper, it was concluded that the gene duplications events within pepper occurred after divergence from tomato. Linkage distances of conserved gene orders were quite similar.

## PHYSICAL AND GENETIC DISTANCES

All distances found on a linkage maps are the product of genetic recombination and therefore are considered to be genetic distances. Two pairs of loci that are genetically the same distance apart may not be physically the same distance apart because of suppressed recombination in the region between two of the loci. Those loci where suppression of recombination does occur will physically be farther apart. RFLP hybridizations to large fragments of DNA are necessary to accurately gauge the physical distance.These experiments are similar to other Southern hybridizations.

The DNA is cut with restriction enzymes first. The choice of enzymes is important because you want to generate large fragments. For these experiments you want to cut the DNA with enzymes that recognize 8 nucleotide sequences in the target DNA. Because these sites will be rare (every 65,536 bases on average) the digestion products will range from several hundred kilobase to several megabases. Because the normal gel systems do not separate large fragments very easily, it is necessary to run pulsed-field-gel-electrophoresis. This technique was first applied to the separation of yeast chromosomes and was quickly applied to large-scale mapping experiments.

Physical and genetic distances are then resolved by hybridizing clones which define closely linked genetic markers, to DNA that has been cut with rare cutting enzymes. What you are searching for is co-hybridization of the two clones to the same restriction fragment. If two clones hybridize to the same fragment, then the maximum distance between those two clones is the size of the restriction fragment. Because you already know the genetic distance between the loci, you can correlate the genetic and physical distances in this region. The following table give the relationship between the physical and genetic distance in three species.

| Species | kilobases/centimorgan |
|---|---|
| Arabidopsis | 139 |
| Tomato | 510 |
| Corn | 2140 |

But remember, these values are estimates of a single region of the genome of each species, and it is quite common to see that two regions of the same species have different amount of DNA per genetic distance. Correlations between physical and genetic distances have also been performed using deletion stocks in wheat.

A series of deletions stocks of a specific wheat chromosome were analysed with a probes known to hybridize to that chromosome. "Each locus was assigned to the chormosome region between the breakpoint of the largest deletion where the band was present and the next larger deletion where the band was absent." (Werner *et al.* PNAS 89:11307–11311) The authors noted large discprencies between the physical and genetic maps of chromosomes 7B and 7D. Two loci that are located near the centromere of 7B are 7 cM apart genetically, yet the distance between the two loci spans 25% of the chromosome. Another region at the distal region of long arm of 7B which accounts for about 15 % of the chromosome is 91 cM long genetically. This points out the difference of recombination that can occur within a single chromosome.

## LINKAGE DRAG

One goal of plant breeding is to introduce a gene from a donor parent to improve a cultivar for a specific trait. For example, a wild germplasm could be used as a source of disease resistance. At the same time the breeder does not want to carry any of the other genes from the wild germplasm that might reduce the a agronomic fitness of the cultivar.

The backcross method of plant breeding is one manner in which the introduction of a specific gene is accomplished. One genetic feature though of backcross breeding is linkage drag. This refers to the reduction in fitness in a cultivar due to deleterious genes introduced along with the beneficial gene during backcrossing. Molecular makers offers a tool in which the amount of wild or alien DNA can be monitored during each backcross generation.

First, though, lets look at the amount of alien DNA that can be maintained after a backcrossing programme. The accompanying figure handed out in class shows the chromosomal region around the Tm-22 allele in a number of tomato cultivars. This allele was introduced from the wild tomato species L. peruvianum, and it provides resistance to tobacco mosaic virus. Large variation in the amount of introgressed DNA was observed. For example, Craigella-Tm-22 contains 51 cM of wild DNA, whereas Vendor-Tm-22 and Nova-Tm- 22 each contain about 8 cM of introgressed DNA.

Backcrossed lines could be developed rapidly in which only a small amount of foreign DNA is linked to the gene of interest. After the first backcross all lines with the gene of interest could be screened with an RFLP that is 1 cM away. Those lines in which a crossover occurred at this marker would be selected. Then those lines would be crossed to the recurrent parent.

The progeny from these crosses would then be scored for a second marker 1 cM away on the other side of the gene. Again crossovers progeny could be selected. Thus, in only two generations the amount of wild DNA could be reduced to 2 cM. The key to this procedure is to have markers that are polymorphic between the two parents that are closely linked to the gene of interest. Young and Tanklsley applied these principles to the reduction of L. peruvianum DNA in Craigella-

Tm-22. The authors were able to reduce the amount of DNA on one side of the Tm-22 allele from 47 to 7 cM in one generation by selecting for crossovers at the CD32A locus.

## QUANTITATIVE TRAIT LOCUS

Traditional quantitative genetic research defined a quantitative trait in terms of variances. The total phenotypic was first partitioned into genetic and environmental variances. The genetic variance could then be further divided into additive, dominance and epistatic effects. From this information it was then possible to estimate the heritability of the trait and predict the response of the trait to selection. It was also possible to estimate the minimum number of genes which controlled the trait.

Mapping markers linked to QTLs identifies regions of the genome that may contain genes involved in the expression of the quantitative trait. But what functions could these genes be encoding. To answer this question we should consider a trait such as yield. What types of qualitative genes (genes inherited as simple genetic factors) could be involved in the expression of yield? The first event required for yield is meiosis.

Therefore any gene that is involved in gamete formation could potentially be considered a QTL. Any of the genes involved in the protein and carbohydrate biosynthetic pathways could also affect the final yield of a plant and could also be considered to be QTLs. As we saw above, the markers associated with a QTL each account for only a portion of the genetic variance. Likewise each of these genes of known function may only account for a portion of the final yield. An important question that can now be posed is whether any known genes map as QTLs.

Beavis *et al.* analysed four populations of maize and found molecular markers linked to plant height. No marker was consistently associated as a QTL with plant height in all four populations. Each of the ten maize chromosomes contained a marker linked to a QTL for at least one of the four populations. The authors further were able to demonstrate that a number of the QTLs identified by the molecular markers mapped to regions containing genes known to have a qualitative effect on plant height.

For example, on chromosome 9 the gene d3 resides within 10 cM of a plant height QTL. This gene is involved in gibberellic acid sensitivity. Mutants do not respond to the hormone and do not undergo the normal cell elongation. These mutants are phenotypically shorter than normal maize plants. The question that needs to be raised is if the QTL that was being identified by the molecular marker is actually the d3 gene. It could be possible that what is actually being measured by the marker is the linkage of the marker with the gene. The statistical analysis of quantitative traits provided valuable information for the plant breeder. Molecular analysis of quantitative traits now provides new tools, not only as selection tools for plant breeding, but as starting points for the cloning of these genes. These objectives could not have been realized without molecular markers.

## PLANT GENOME ORGANIZATION AND STRUCTURE

Eukaryotic genomes are much more complex than prokaryotic genomes. And further, plant genomes are more complex than other eukaryotic genomes. Prior to the development of recombinant DNA technology genomes, were analysed by reassociation kinetics techniques. Reassociation kinetic experiments are performed by melting DNA and allowing it to reanneal upon itself or with another population of either DNA or RNA molecules. The kinetics of the reassociation provide data that can be used to analyse the overall structure, evolution and expression of genomes.

The most common method of denaturing duplex DNA is by heating to 100°C. But how can we monitor this denaturation and subsequent renaturation? The most common method is by measuring the absorbance change in the ultraviolet region at 260 nm. The important property is that melted, single-stranded DNA absorbs about 40% more at 260 nm than duplex DNA. If we slowly heat DNA the absorbance will increase dramatically over a short range of temperatures. The mid-point of this transition is called the melting temperature or Tm. Under physiological conditions, the Tm usually lies in the range of 85-95°C. Thus, without altering the cellular conditions, the duplex DNA is stable in the cell. The exact temperature that a particular DNA melts depends on several parameters. The GC content is important because GC base pairs have three hydrogen bonds compared to the two for AT base pairs. Thus, the higher the GC content the higher the Tm.

**Table. Effect of GC Content on Tm**

| %GC | Melting Temperature |
|---|---|
| 40 | 87°C |
| 60 | 95°C |

DNA denaturation is reversible, and the reversible process is called renaturation. This process requires that the temperature be lowered gradually.

*As this lowering of temperature proceeds the following occurs*:

- Single-stranded molecules randomly encounter each other.
- Short complementary stretches of duplex DNA form.
- The DNA then zips back together to form the original structure.

Obviously if a single molecule is denatured it should be able to reform completely. But if two separate molecules are denatured together, for example wheat and barley DNA, then complementary regions between the two DNAs will be able to form duplex DNA. The ability of two molecules to renature is called hybridization.

*Hybridization*: The pairing of complementary nucleic acids. Performing hybridization experiments in a solution is called liquid hybridization. We have already discussed the related procedure called filter hybridization.

## ANALYSIS OF GENOMES BY REASSOCIATION EXPERIMENTS

If a DNA molecule is melted and allowed to reassociate, the complexity of the genome dictates the rate in which duplex DNA will form. If we consider a simple molecule that consists of alternating GCs, this molecule will be able to form a duplex quicker than a molecule that consists of repeating blocks of AGCT. As the number of different combinations of bases increases, the time required for complete duplex formation to occur will increase.

Renaturation, or duplex formation requires random collisions between two single-stranded molecules. This process follows second-order kinetics and is concentration dependent. We will not go through the derivation of the formula but the important parameter used to define a certain DNA is:Cot½.

This value is defined as the amount of time required for one-half of the DNA to reanneal or form duplex DNA. The units for this parameter is moles of nucleotides per liter per second. The more complex the genome of interest, the longer it will take for like sequences to reanneal. Consequently, the Cot½ will be larger. Thus in terms of reassociation kinetics complexity has a specific definition.

*Complexity*: The total length of different sequences. For example, E. coli is considered to have a complexity of 4.2 × 106 base pairs. What is the experimental procedure used to derive these values?

*In general the procedure is*:

- Shear the DNA to be analysed to a length of about 300 bp.
- Melt the DNA (usually in 0.12 M phosphate buffer) by boiling for 5 min.
- Quickly place at 60°C.

- Take aliquots at different time points. Separate single-stranded DNA from double-stranded DNA by hydroxyapatite. Measure the amount of DNA that is double-stranded by absorbance at 260 nm.
- Plot the amount that is single-stranded versus the Cot value. The Cot value is expressed in log equivalent. This plot depicts the Cot curve.

When this type of experiment is performed with eukaryotic DNA three components are usually seen. These components each reanneal with their own unique Cot½ value. The three components are termed the fast, intermediate, and slow components. Why do we see these three components? Eukaryotic genomes are characterized by sequences that are represented by different copy numbers. If a sequence is found many times in the genome, it will reanneal much quicker than those sequences that are found only once in the same genome. Thus the equivalent Cot curve for a eukaryotic genome will be different than a genome, such as E. coli, which only contains single copy sequences.

A comparison of the Cot value of each of these components with an E. coli standard allows us to derive the complexity of each component. The complexity of the slow component of the genome is greater than that for the other two components and is considered to represent the single copy portion of the genome. The complexity of the slow component can be used as a good estimate of the genome size. The genome size will be the sum of the lengths of all the unique sequences. Using the example from Genes V - Lewin, p.664, the complexity of the slow component is $3 \times 10^8$ bp and the complexity of the intermediate component is $6 \times 10^5$. If we divide the complexity of the slow component into the intermediate component we get $2 \times 10^{-3}$. This demonstrates that the intermediate component contributes very little to the complexity of the genome. Therefore, the complexity of the slow or single copy portion of the genome can be considered equal to the genome size.

To derive the complexity of each component, it is necessary to run a standard, such as E. coli DNA, with each experiment. As stated above, E. coli is considered to consist of only single- copy sequences. Let's say that in the experiment from which the Cot for each component was derived, the Cot½ value for E. coli was 4. Experimentally it was determined that the slow component comprised 45% of the total DNA. Therefore, if only that component was annealed, the Cot½ value would be 283 ($630 \times 0.45$). That value is 71 (283/4) times slower than for E. coli. Therefore the complexity of the slow component is 71 times that of E. coli or $3.0 \times 10^8$ ($71 \times 4.2 \times 10^6$). The complexity of the other components is derived similarly. Genome size is quite variable throughout the biological world and the genome size in plants shows the greatest variation of any kingdom in the biological world.

**Table. Variation in Genome Size among Plants**

| Species | kb/haploid | pg/haploid |
|---|---|---|
| Arabidopsis | $7 \times 10^4$ | 0.15 |
| Lily | $1 \times 10^8$ | 100.00 |

*Conversion factor*: 1 pg = 0.965 × 109 bp = $6.1 \times 10^{11}$ daltons

One manner in which a genome can be described is by determining the distribution of fast, intermediate and slow components in the genome. For comparison purposes, what does the human genome look like? Distribution Sizes Among Components of the Human Genome

| Component | % of Genome |
|---|---|
| Fast | 6 |
| Intermediate | 38 |
| Slow | 50 |

The table Sequence Distribution of Selected Plant Species lists the different components from different plant species. As you can see, plant species exhibit a wide range of values for each of the components. The genome of Arabidopsis is essentially entirely single copy sequences (the repetitive sequences have been determined to be essentially all chloroplast DNA).

At the other extreme, pea and wheat genomes have only 10–20% single copy sequences. Reassociation kinetic experiments of polyploid species, such as bread wheat (Triticum aestivium ) were unable to derive a component that displayed true single copy kinetics. Instead the slowest component appeared to act as if it consisted of copies represented three times.

This result is consistent with the current hypothesis that bread wheat was developed from the introgression of three diploid wheat species. Genomic analysis suggest that this hypothesis is correct since the slowest component appears to consist of sequences represented three times.

## REPEATED SEQUENCES

The intermediate and fast components are composed of sequences that are found many times in the genome. These sequences are called repetitive sequences and can vary in size from a 100 bp to 1000 bp or more. Furthermore, these sequences have undergone sequence divergence by the addition or deletion of sequences or by changes in the base pair sequence. Thus, the repeated sequences themselves show some divergence. An example of a highly repetitive sequence is the repeat found to be associated with the knob heterochromatin of corn. It ranges from 3-5 × $10^5$ copies on a small knob to 1 × $10^6$ on the large knobs. This sequence is unique to knobs and is not found associated with any other heterochromatic regions of corn. An example of a functional repeated sequence in plants is the corn storage proteins, zeins. Two major classes of zeins exist, the 22 and 19 kd classes. Sequence analysis has shown that both classes have the same structure.

The only difference between the 22 and 19 kd class is the repeat unit. The 22 kd class has 8 repeat units and the 19 kd class has 7 repeat units. Estimates have been made of the number of copies of these genes and 30-50 copies of the 22 kd class are found in the corn genome. Thus, the repeat unit would be represented 240-400 times in the genome.

## ORGANIZATION OF SINGLE-COPY SEQUENCES

Single-copy sequences are interspersed throughout the plant genome. These sequences are bounded by repeat sequences. The length of the single-copy regions varies widely among plant species.

*In general, two types of arrangements are recognized*:

- *Short Period Interspersion*: Single copy sequences of 300-1200 bp are interspersed as islands among short lengths of repeat sequences
- *Long Period Interspersion*: Single copy sequences of 2000-6000 bp are interspersed as islands among repeat sequences

How can the interspersion type be determined? As we discussed earlier, reassociation kinetic experiments are performed with 300 nt long fragments. These experiments give a characteristic Cot curve that defines each of the components. But what happens if the fragment that is followed is of a longer length, for example 900 nt? This fragment could be considered to consist of three 300 nt fragments, and each fragment may be from any of the three components.

Indeed this is how these experiments may be performed. A tracer of a longer length (such as 900 nt) is prepared and radiolabelled, for example with $H^3$. This tracer is added to a normal Cot reaction in such a low concentration that the rate is not affected. The reaction is then allowed to proceed to a Cot value determined from a normal experiment with only 300 nt fragments where the repetitive sequences have reannealed, but the single-copy sequences have not.

Then the amount of tracer which remains single-stranded is determined. What will be seen is a reduction in the amount of expected single-copy sequences. Why? If the 900 nt long fragment contains both single copy and repetitive sequences it will reanneal under these conditions as a repetitive sequence, and thus the amount of DNA that appears in the single-copy fraction will be reduced proportionally. Therefore, that single-copy sequence is interspersed with repetitive sequences. If for example the single copy fraction is determined to account for 40% of the genome when 300 nt fragments are used, and we calculate that the genome has 30% single-copy sequences when a 900 nt tracer is used, 10% (40%-30%) of the 900 nt fragments contain single-copy and repeat fragments and that 25% (10%/40%) of the single-copy DNA is interspersed with repetitive sequences at an interval of 300-900 nt.

A final concern is the size of the repeat units. This can be obtained by using a long tracer (5000 nt, for example) and reannealing to an appropriate Cot such that only repeat sequences bind to the tracer. The product is then treated with an enzyme that cuts only single-stranded DNA. This enzyme is S1 nuclease. After digestion you obtain products that only contain repeat sequences. These products are then sized to give the average, mode and range of repeat sequences.

## EVOLUTION OF REPEATED SEQUENCES IN CEREALS

The analysis of repeated sequences has also provided experimental evidence which supports the current model of cereal speciation. The following tables and the associated diagram detail the results. These show that once a repeat enters the lineage it remains. Therefore group I repeats are found in all of the species because it is the most ancient repeat. Secondly, new repeats are added to the lineage and appear in all species which subsequently diverged from the lineage. Finally, species-specific repeats are formed in each species after that species has diverged from the common lineage. The distribution of the lineage specific repeat supports the speciation model.

## ESTIMATING THE NUMBER OF EXPRESSED GENES

Reassociation kinetics can also estimate the number and abundance of expressed genes. These experiments are performed using high concentrations of RNA and tracer levels of either DNA or cDNA. These experiments are analagous to DNA reassociation experiments except the units are expressed as moles of ribonucleotides per liter per second and the value is called a Rot. If tracer amounts of DNA are allowed to reanneal only a portion of that DNA will form a duplex because not all the DNA will be expressed in the RNA population to which it is being hybridized. Let's use the example in the Genes V to derive an estimate for the number of genes that are being expressed. In this experiment 1.35% of the DNA hybridized to the RNA. Since RNA is single stranded the other anti-sense strand of the DNA would not have a partner with which it could hybridize. Thus actually 2.7% of the DNA is represented in this RNA population. If the genome size is 8.1 X 108 bp and the single-copy sequences represent 75% of the genome then we can estimate the complexity of the expressed DNA.

$0.027 \times 0.75 \times ( 8.1 \times 108$ bp $) = 1.7 \times 107$ bp If each gene is about 2000 base pair then the number of genes that is expressed is:

$$7 \times 107 \text{ bp} / 2000 = 8500 \text{ genes}$$

## CHLOROPLAST GENOME ORGANIZATION

All angiosperms and land plants have cpDNAs which range in size from 120-160 kb; three expceptions are:

| Species | Size (kb) |
|---|---|
| N. accuminati | 171 |
| Duckweed | 180 |
| Geranium | 217 |

All cpDNA molecules are circular and spinach is used as the basis for all comparisons. Very few repeat elements are found other than short sequences of less than 100 bp. The notable exception is a large (10-76 kb) inverted repeat section, which when present, always contains the rRNA genes. (Legumes such as pea do not contain this repeat.) For the majority of species, this repeat region is 22-26 kb in size. Finally,the genetic order of the ribosomal unit is conserved in all species:

$$16S - tRNA^{ile} - tRNA^{ala} - 23S - 5S$$

Two other features of chloroplast DNA is described. First it was shown to that it can exist in in two orientations This implies that the molecule can undergo an isomerization event. Second is has been shown that spinach, corn, tomato and pea can all exist as multimers (PNAS 86:4156, June 1989).

| Multimer | Relative Abundance | Per cent |
|---|---|---|
| Monomer | 1 | 67.5 |
| Dimer | 1/3 | 22.5 |
| Trimer | 1/9 | 7.5 |
| Tetramer | 1/27 | 2.5 |

Because photosysnthesis is the primary function of the chloroplast it is not surprising that the chlroplast genome contains genes which encode for proteins that are involved in that process.

| Reaction | Function |
|---|---|
| Dark Reactions | rbcS (nuclear encoded) |
| | rbcL (chloroplast encoded) |
| Light Reactions | apoproteins for PSI andPSII |
| | cytochrome b6 |
| | cytochrome f |
| | 6 of 9 ATPase subunits |
| | cab, LHC proteins (nuclear encoded) |
| | plastocyanin (nuclear encoded) |
| | ferredoxin (nuclear encoded) |
| Other | 19/60 ribosome binding proteins |
| | translation factors |
| | RNA polymerase subunits |
| | tRNA and rRNA genes |

Atrazine resistance is apparantley mediated through the psbA gene sequences of the 32 kd protein which is encoded by cpDNA. DNA sequence analysis revealed the following amino acid changes that are thought to be important.

| Species | AA# | Susceptible | Resistant |
|---|---|---|---|
| Blue green algae | 264 | Ser (TCG) | Ala (GCG) |
| Chlamydomonas | 264 | Ser (TCT) | Ala (GCT) |
| Solanum nigrum | 264 | Ser (AGT) | Gly (GGT) |
| Amaranthus | 228 | Ser (AGT) | Gly (GGT) |

## Evolutionary Changes of cpDNA

- The majority of changes are small insertions and deletions of 1-106bp; significantly, a few length mutations of 50-1200 bp are clusted in "hot spots".
- The largest deletion occured in pea where an entire rRNA cluster is lost.
- The most common evolutionary change is in gene order. Small changes in the gene order occur, especially in the algae, but inversions have generated large scale order changes:
  - *Legumes*: About 50 kb inversion brought rbcL closer to psbA
  - *Wheat*: About 25 kb inversion brought atpA closer to rbcL

## MITOCHONDRIAL GENOME ORGANIZATION

In comparison to the chloroplast genome, the size of the mitochondrial genome is quite variable.

| Species | Size (kb) |
|---|---|
| Oenothera | 195 |
| Turnip | 218 |
| Corn | 570 |
| Muskmelon | 2400 |

Further, in comparison to the mitochondrial genomes of other species the size is quite large and variable. For example, animal mitochondrial genomes range in size form 15-18 kb, and fungi mitochondrial genomes range form 18-78 kb.

Plants may code for more proteins than with species. For example, genes for ribosomes, subunits I and II of cytochrome oxidase and ATPase subunits are located on the mitochondrial genomes of plants.

When DNA from corn mitochondria was investigated with EM, several circular molecules of different sizes were detected. Once the genome was mapped it became apparent that a mechanism existed to generated these circles of different sizes. It is now understood how these molecules arise. First, lets look at the simple situation of turnip. Two direct repeats undergo intramolecular recombination to give the two smaller molecules:

218 kb → 135 kb + 83 kb

The mitochondrial genome of corn undergoes the same type of recombination, but the events are more complex. First, the master circles can be subdivided into two major subgroups:

570 kb → 488 kb + 82 kb

570 kb → 503 kb + 67 kb

The second group of molecules are still labile and can produce several other subpopulations. Further two subgenomic circles can unite to form a larger circle. This variability is possible because corn has 10 repeats with which intramolecular recombination can occur.

| Species | Master Circle Size (kb) | Sub-genomic Circle Size (kb) | Repeat Size (kb) |
|---|---|---|---|
| Turnip | 218 | 135 + 83 | 2 |
| Cauliflower | 217 | 172 + 45 | ? |
| Black Mustard | 231 | 135 + 96 | 7 |
| White Mustard | 208 | none | none |
| Radish | 242 | 139 + 103 | 10 |
| Spinach | 327 | 234 + 93 | 6 |

Introns have been located in the cytochrome oxidase subunit II gene (the cytochrome complex consists of 3 mitochondrial and 4 nuclear encoded genes). This gene contains one intron in rye, corn, wheat, rice, and carrot, but for other species such as Oenothera, broad bean, cucumber the gene has no intron.

## Promiscuous DNA

Stern and Lonsdale (1982) hybridized mtRNA to a SstII digest of maize mt DNA and found that it hybridized to fragments known not to contain mt rRNA genes. The question of interest was - what was it hybridizing to? They next looked at a cosmid clone of corn mtDNA that hybridized to the mt RNA and found that it hybridized to a RNA molecule of the size of the cp 16S RNA gene. How could this have happened?

They next mapped the clone and compared it to the map of the corn cpDNA and found that the clone map was almost congruent with that of the the cpDNA 16S RNA region. This mapping showed that the two maps were nearly identical over a 12 kb region of DNA. These results suggest that cpDNA had been transferred to the mitochondrial genome.

The observation that organelle DNA was found in other DNA compartments of the cell was extended by other researcher. Stern and Palmer looked at corn, mung bean, spinach and pea and found extensive evidence of cpDNA/mtDNA homology. These observations were extended to other DNA locations in the plant cell. Kemble *et al* (1983) demonstrated that mitochondrial DNA sequences are located in the nucleus of corn. Scott and Timmis (1984) showed that cpDNA sequences are found in the nuclear DNA.

## RNA EDITING

The Central Dogma of Molecular Genetics states that the information that is found in DNA is used to produce mRNA molecules that are instrumental in the production of proteins. Therefore, the information flows directly from DNA to protein, via the RNA intermediate molecule. Recently it has been discovered that the information that is contained in the DNA is not always found in the RNA products used to make proteins.

It has now been demonstrated that mitochondria and chloroplast contain the biochemcial machinery to alter the sequence of the final transcription product. This process is called RNA editing. This process was identified in the following manner. Sequence analysis of a number of cytochrome c oxidase subunit II genes from non-plant species revealed that a tryptophan residue was invariant at several locations in the final protein product. But sequence analysis of this gene in several plant species revealed arginine at those positions.

Since a single base pair change in the codons for the two amino acids could generate this change (CGG for UGG), it was suggested that CGG encoded for tryptophan and not arginine in plant mitochondria. (This is the only change in codon usage that has been suggested for plants and has been postulated for several other genes as well.) But this change in codon usage was not universal, that is some CGG codons actually specified arginine in the final protein product. Furthermore, no amino acyl tRNA that recognized CGG was found to be charged with tryptophan, a prerequisite if this codon specification was actually real.

The solution to this dilemma was found by sequencing the mRNA products for cytochrome oxidase subunit II genes. It was found that in the mRNA the cytosine residue had been changed (edited) to uridine at the sequence location where the invariant tryptophan residue is found. This changed the codon at that location to UGG which is recognized by a tRNA that carries the amino acid tryptophan. An analysis of three other plant mitochondrial genes where the same altered codon usage was predicted suggested that mRNA editing was also occurring at the codon and that a cytosine residue was edited to uridine.

This editing process has also been detected in protozoa and it remains to be determined if RNA editing is a widespread function in mitochondria. A final point that this editing function highlights is that the sequence that is found in the DNA is not entirely and faithfully represented in the final protein product.

## Specific Features of RNA Editing

- Editing can occur in both mitochondria and chlorplasts
- To date, >300 different editing events have been detected in plant mitochondria.
- The vast majority of the events involve a C to U tansition. A few cases of U to C transitions have been reported. This suggests that the editing machnery can also carry out the reverse modification.
- RNA editing can modify from 0.8% to 5.8% of the nucleotides of a specific transcript.
- The RNA editing events appear to occur at random in the transcript.
- Both 5' and 3' non-coding regions of mRNAs have also been shown to be edited.
- Structural RNAs such as tRNAs and rRNAs do appear to be affeceted.

- Editing can convert a tryptophan codon to a arginine codon (CGG to UGG).
- Start AUG codons can be created from ACG threonine codons
- Stop codons can be created by editing CAG, CAA and CGA codons.
- The most frequent amino acid substitions derived from RNA editing are Pro to Leu, Ser to Leu and Ser to Phe.
- Plant mitochondria do not use the universal genetic code.

The primary benefit of RNA editing could be evolutionary conservation of protein structure. For example, bound copper is required for the funciton of cytochrome c oxidase subunit II (coxII). After editing, all amino acids #228 are converted to cysteine, an amino acid required for copper to bind. In all species except for plants, the coxII gene encodes for methionine at codon #235. In plants, this methionine is generated by RNA editing. These events suggest that this protein is under very strong structural and functional constraints.

# Chapter 12

# Functional Genomics

## PLANT CELL BIOLOGY

- *Plants are photoautotrophs*: The most obvious distinguishing characteristic of plants is the presence of chloroplasts, which enable them to photosynthesize. They are thus photoautotrophs. The presence of plastids also has implications for genetics, since plastid genomes can carry important traits, most of which are directly related to photosynthesis.
- *Cell walls constrain development*: Plants have cell walls made of cellulose, fungi have chitinous cell walls, while animals do not have cell walls. Plants maintain their structural integrity through a combination of mechanisms: turgor, cell walls, and insoluble components such as lignin. Animals derive their structural integrity through endo- or exo-skeletons, muscle and connective tissue. Higher plants and animals have circulatory systems that allow for the distribution of nutrients and the removal of wastes. Fungi are constrained in size due to the lack of any comparable features.
- Another consequence of the cell wall has to do with the limitations it imposes on diffusion. With pores of 3.5 to 5.2nm, movement of molecules >15Kd is significantly impeded. Thus, plants must subsist on molecules of low molecular weight, and any intercellular signaling molecules that have to pass through the cell wall must also be small.
- *Plasmodesmata facilitate symplasty*: Plasmodesmata convert a plant from being just a collection of individual cells to a large, interconnected symplast. Although the passage of molecules through plasmodesmata is tightly controlled, this provides a means for intercellular communication that is not available in animal tissues.
- *Most cell volume is taken up by vacuoles*: Often 50%, to as much as 95% of the volume of a plant cell is taken up by vacuoles. The

vacoule makes it possible for plant cells to become very large, without a correspondingly large increase in cytoplasmic volume. This is important because the cytoplasm is requires a fairly large investment, in terms of macromolecular synthesis and turnover. Additionally, by limiting the volume of the cytoplasm, the rates of intracellular diffusion are not slowed down. This in turn affects diffusion-dependent processes, such as signal transduction from the plasmalemma or cytoplasm to the nucleus, protein synthesis, protein transport, membrane synthesis, cytoplasmic streaming, and absorption and equilibration of nutrients, gases, and other small molecules.

## LIFE HISTORY OF PLANTS

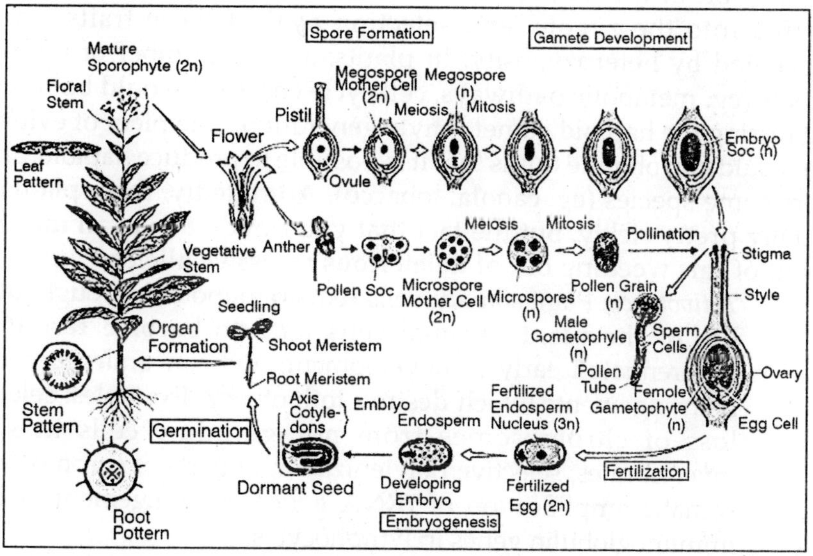


**Fig.** Life Cycle of Flowering Plant

- *The meristem (germline)*: The germ line - Animals have specialized germline tissues, spermatogonia and oogonia, which diverge from somatic cells early in embryogenesis, and are largely insulated from selective pressures. In contrast, flowering plants have large portions of their undifferentiated tissue as meristem, tissue capable of growing vegetatively and differentiating (often in response to very specific environmental cues, such as day length) into reproductive organs which then produce germ cells.
- *Gametophyte and sporophyte generations*: In animals, the final meiotic products (spermatids and ova) are mostly metabolically dormant, and actually shut down gene expression when

they have completed maturation. In fact, egg cells in animals typically have enough mRNA and rRNA pre-synthesized to last the zygote for several cell divisions after fertilization.

Flowering plants have a 2 part life cycle, a diploid (or greater) sporophyte stage, and a haploid gametophyte stage.

*Female*: Megaspore mother cell (2N) → megaspore (1N) → megagametophyte → egg

*Male*: Microspore mother cell (2N) → microspore (1N) → pollen grain → sperm

Thus plant germline (gametophyte) cells undergo a number of cell divisions in the haploid state, and, presumably, have to perform all of the normal metabolic housekeeping functions. An important implication of this is that recessive lethal alleles in animals will be carried into the zygote, and selection against these traits may be prevented by heterozygosity. In plants, a good number of recessive lethals (eg. metabolic pathways, cell cycle controls) would be weeded out during the haploid gametophyte generation. One piece of evidence that could corroborate this is that it is possible to produce haploid plants from some species (eg. canola, tobacco). Admittedly, these plants are usually pretty sickly, but the fact that you can do this at all may be a result of this weeding out of deleterious recessive alleles.

- *Totipotency*: Plant somatic cells remain totipotent throughout the life of the plant. Animal cells tend to become terminally differentiated early in development, and can undergo genetic rearrangements which destroy totipotency. Examples: selective loss of chromosomes from non-germline cells in some invertebrates; selective polytenization of parts of chromosomes; somatic amplification of rRNA genes; rearrangements within immunoglobulin genes in lymphocytes.

*To elaborate on this last example*: In vertebrates, the immunoglobulin genes in lymphocytes undergo rearrangements of DNA in the regions encoding the hypervariable regions of antibody proteins. Since a different rearrangement occurs in each cell, the population of lymphocytes can differentiate into T-cells and B-cells, each of which will carry antibodies with unique specificties. These DNA rearrangements are irreversible. The totipotency of plant cells has important implications. First, it deprives plants of one, admittedly drastic, means of genetic regulation. More importantly, it has made the transformation of plants for experimental and genetic engineering purposes a great deal easier.

- *Differentiation without migration*: Animal cells must migrate by amoeboid motion through the embryo during development.

Plant cells are constrained by their cell walls to stay where they are. Therefore, differentiation must be carried out without migration. This implies that there are very fundamental differences between plant and animal ontogeny. Is this inability of cells to migrate an evolutionary dead end for plants? Is this the key to why animals seem to have a greater potential for physiological complexity than plants?

- *Continuous programme of development*: The totipotency of plant cells, along with the relative lack of terminal differentiation during vegetative growth results in a continuous programme of differentiation and development throughout the lifetime of the plant. In animals, once an organ or tissue has formed, that's it. You could say that, while animals go through a single iteration of the developmental programme, plants can go through multiple iterations, usually in response to environmental stimuli. This can be viewed in its extreme as the abilty of new plants to be produced by vegetative propogation (eg. potatoes) without ever going through a gametophyte stage.
- *Shift between vegetative and reproductive growth*: In animals, germline cells become differentiated very early in development, and specialized sex organs are established early. In flowering plants, sporophytes grow vetatively for an extended period. The shift from vegetative to reproductive growth is triggered by environmental cues. It is during flowering that reproductive organs develop.
- *Kinetics of cell growth*: The same kinetic principles that govern bacterial cultures govern plant growth. In a sense, cells in an organ can be thought of as a culture.

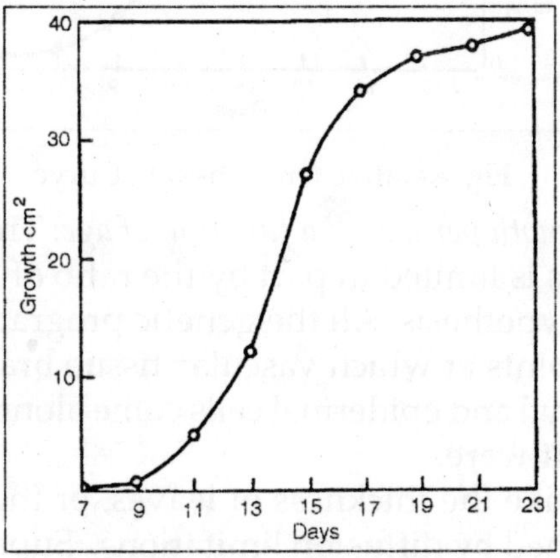


**Fig.** Sigmoid Growth Curve

Note that, as in bacterial growth, growth goes through an initial lag phase, followed by a phase of logarithmic growth, followed by a stationary phase.

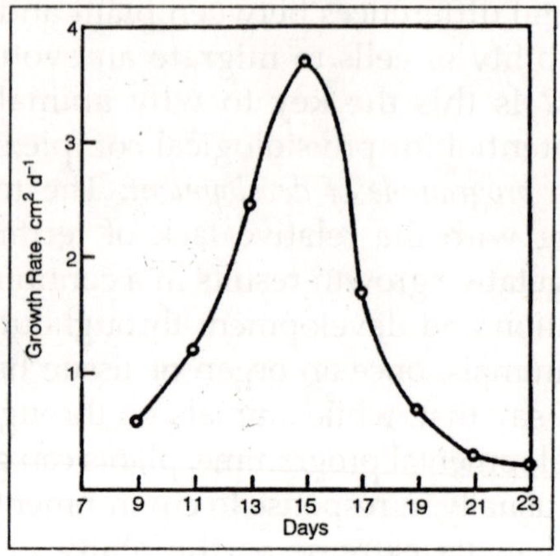


**Fig.** Growth- Rate Curve

The first derivative of the curve above, simply indicating that the growth rate reaches a maximum, and then tapers off.

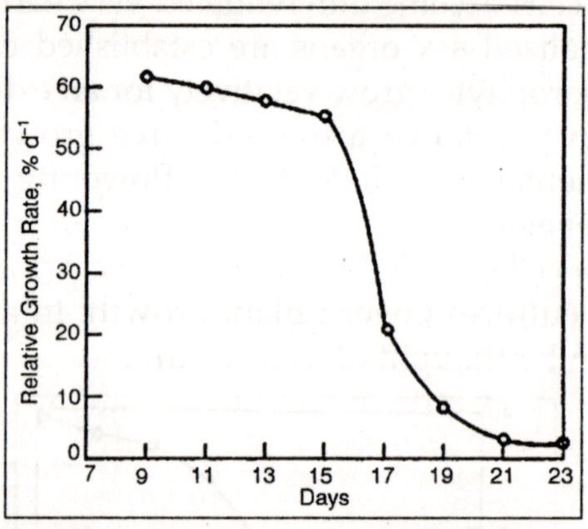


**Fig.** Relative Growth-Rate Curve

*Per centage growth per day as a function of age*: The growth of a local population of cells is limited in part by the ratio of the mass of cells to vascular tissue. Hypothesis: All the genetic programme needs to do is control the key points at which vascular tissue branches out, and the adjoining mesophyll and epidermal cells come along for the ride, filling out the shape, as it were.

Thus, things like the thickness of leaves, or the distance between veins are determined by diffusion limitations. Superficially, we're not just talking about growth as cell division, but there is also a component of elongation. Both these processes must be under the control of developmental regulators.

- *Developmental plasticity*: The plant body plan is more of an iterative loop, rather than a specific blueprint, as in animals. The specific location, and number or organs, is often environmentally determined:
  - Switch from vegetative to reproductive growth (flowering) does not necessarily occur at a specific stage of growth
  - Number of leaves or nodes
  - Tissues such as roots can grow in direction of nutrients; leaves and stems can grow towards light.
- *Localized senescence*:
  - leaves senesce and abscise, while branches persist. (One of the few analogues in animals might be a snake shedding its skin.)
  - biomass from senescent tissue can be resorbed and transported to other tissues eg. seeds.
- *Cell division not required for new cell identities*:
  - Mesophyll cells can differentiate into tracheary elements through a breaking down of the cytoplasm and vacuole, and thickening of the cell walls.

## GENE EXPRESSION

### Global Analysis of Gene Expression

The experiments of Goldberg and others on global analysis of gene expression provided what is, even today, the most comprehensive look at the role of gene expression in development. The key problem with R0t analysis, as done in these studies, is that nothing is learned about which genes or pathways are important in each organ or tissue studied. We can estimate the complexities of the RNA populations, and even learn, for example, that gene expression in leaves is very similar to that in petals, but quite dissimilar to that in roots. However, not knowing the identities of the genes involved limits the usefulness of these findings.

Most of the history of molecular biology has been characterized by the use of cloned probes, typically in Northern blots, to measure gene expression one gene at a time. This is precise and specific, but still has important limitations. First, the number of genes that can be examined is limited for practical reasons. You have to have a clone for every gene you wish to study. Secondly, it is necessary to decide which genes to look at. Unfortunately, choosing which genes to study biases the experiment. Important genes might be missed, and the role of relatively minor genes could be overestimated. One of the unifying themes in the field of genomics is that by working on a large scale, you get a more comprehensive picture of the subject. We will look at several large scale approaches toward finding important genes in biological processes.

### *Gene Arrays*

EST studies begin to tell us about the expression of the genome, but all they really tell us is that a gene is expressed in the cells from which the ESTs were derived. Gene array experiments give a more precise picture of the 'transcriptome', the set of transcribed genes, in specific developmental stages, or in response to treatments.

- *Gene array technology*: Measures mRNA levels for thousands of genes.
  - Any cell or tissue type.
  - At any point in development.
  - In response to any stimulus.

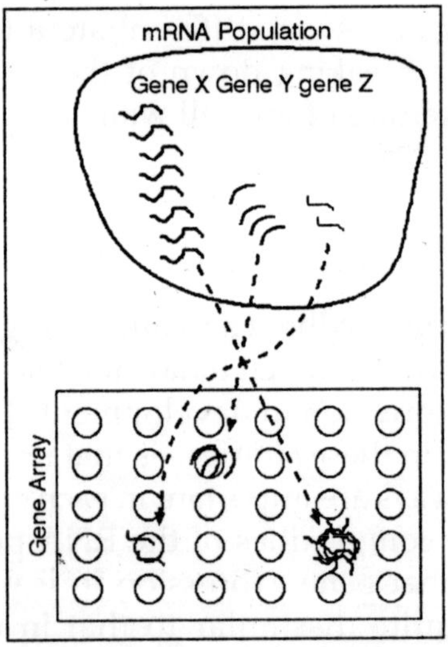


Gene arrays consist of hundreds or thousands of cDNAs spotted onto microscope slides (microarray) or nylon filters (macroarray). cDNAs are chosen from EST collections, so the sequences, and usually the identities of genes in the array are known. In a gene array experiment, an mRNA population is isolated from cells. The population is labeled by synthesizing complementary cDNAs using reverse transcriptase and labeled nucleotides. The resulting cDNA population is then hybridized to the array.

- *Gene* x: Strongly expressed; high abundance transcript
- *Fene* y: moderately expressed; medium abundance transcript
- *Gene* z: weakly expressed; low abundance transcript

Each transcript base pairs with the complementary DNA for its corresponding gene on the array. Signal strength is proportional to the abundance of each mRNA *Example*: cDNA probes were made from Human fibroblasts treated with serum (Cy5-dUTP) or serum-deprived cells (Cy3-dUTP). 3 replicate arrays were scanned, and the same regions from the replicate arrays are shown. Serum inducible genes appear green, while serum suppressible genes appear red. 1 - protein disulfide isomerase-related protein P5; 2 - IL-8 precursor; 3 - EST AA057170; 4- vascular endoghelial growth factor.

- *Two general classes of data*: Gene array studies tend to generate two different types of data. Studies in which two or more conditions are compared at a time generate discrete state data. Often it is critical to follow the expression of a gene over time after a treatment. In timecourse experiments, the expression of each gene in response to two or more treatments is measured over time. For example, in the timecourse at right, the solid blue and red dashed curves might represent the expression levels for a gene in response to two different drugs.

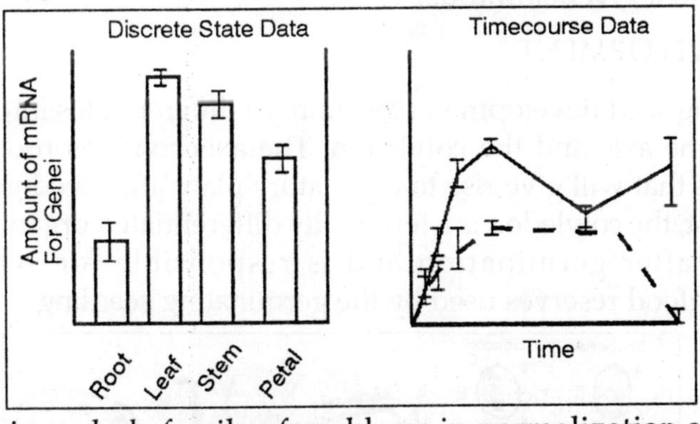


There is a whole family of problems in normalization of data and controlling for components of experimental variation. To put things into perspective, if the experiment was repeated 3 times, the timecourse above represents 2 treatments × 6 times × 3 replicates = 36 probes hybridized to 36 duplicate arrays to generate the data.

Although the data for each replicate are averaged, there is often a great deal of variation in the results, which can potentially negate any meaning. Therefore, extraordinary measures must be taken to minimize experimental variation at each step in the procedure, to minimize the overall variation.

- Gene array technology can generate data for thousands of genes.

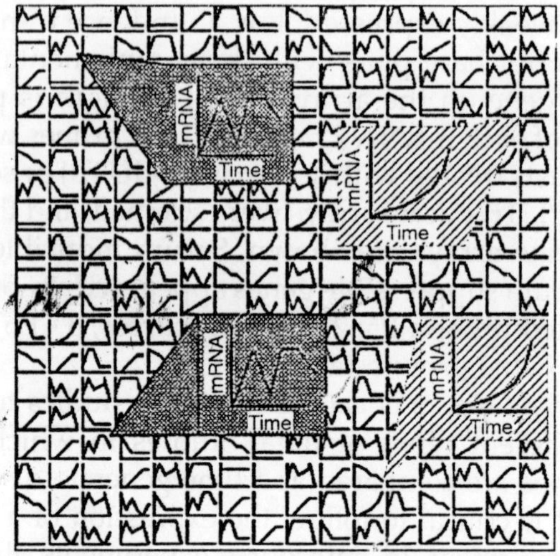


What we're ultimately trying to get from gene array experiments is expression patterns for each of the hundreds or thousands of genes in the array. By identifying genes whose expression patterns are similar, we can discover which groups of genes work in concert, in response to a given stimulus.

## SEED DEVELOPMENT

During seed development, the embryo differentiates into two organ systems: the axis and the cotyledon. The axis contains root and shoot meristems that will give rise to the mature plant after seed germination. By contrast, the cotyledon is a terminally differentiated organ system that senesces after germination and is responsible for synthesizing andstoringfood reserves used by the germinating seedling.

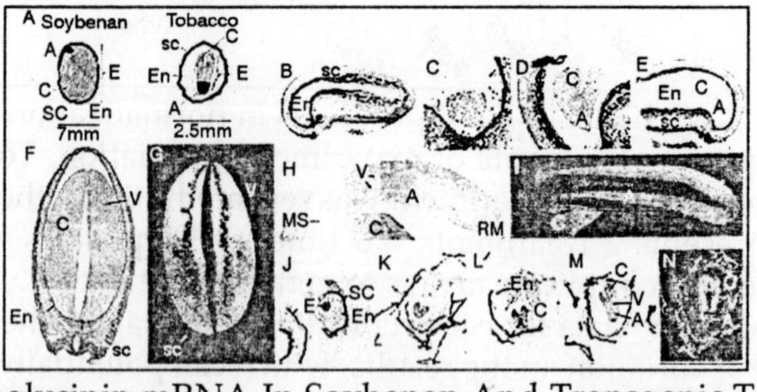


β -Conglycinin mRNA In Soybenan And Transgenic Tobacco.

β-conglycinin - storage protein, accumulates to high levels in soybean seeds. ~15 genes in soybean, clustered in two different chromosomal loci. 2-G - Dark-field microscopy shows silver grains as bright signal against dark background. β-conglycinin mRNA accumulates solely in the cotyledons (c) and not in surrounding embryonic tisssues (e). In the cotyledons, it accumulates in the storage parenchyma tissues and not in the vascular cylinder (v). 2J-M - development of tobacco seed. 2N - In-situ hybridization of transgenic tobacco expressing soybean beta-conglycinin.

- Similar expression patterns to soybean ie. only in cotyledon, non-vascular tissues
- Indicates that signals for tissue-specific expression in developing embryo must be well conserved.

## Embryogenesis

Plant embryogenesis is the process that produces a plant embryo from a fertilised ovule by asymmetric cell division and the differentiation of undifferentiated cells into tissues and organs. It occurs during seed development, when the single-celled zygote undergoes a programmed pattern of cell division resulting in a mature embryo. A similar process continues during the plant's life within the meristems of the stems and roots. Tobacco and soybean embryos contains only about 10 distinct cell and tissue types each, yet the mRNA populations contain roughly 20,000 different transcript classes. This is in contrast to animal development, in which a reduction in the number of genes expressed occurs as embryogenesis proceeds.

### *The Process of Embryogenesis*

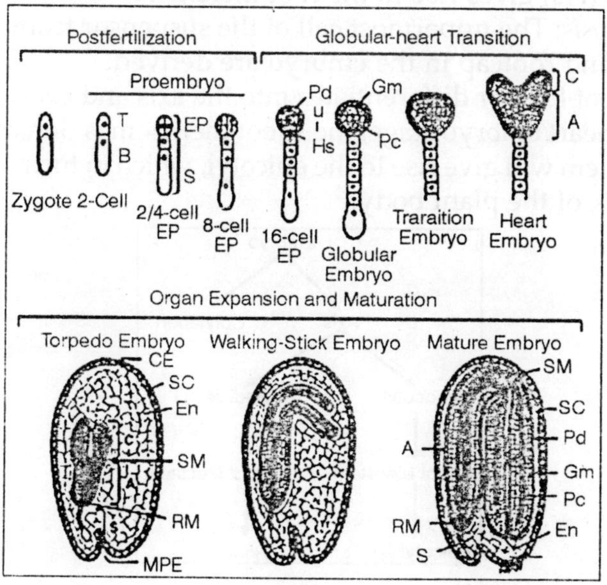

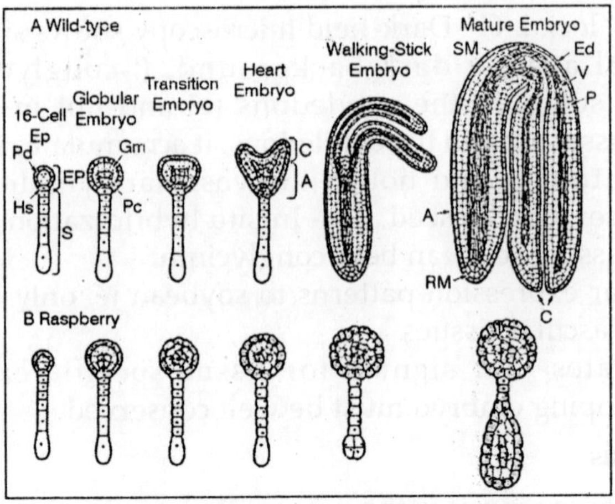


Fig. Overview of Embryogenesis

- Assymetric cleavage of the zygote results in the formation of an embryo with a suspensor and embryo proper.
- *Embryo proper*: Gives rise to embryo.
- *Suspensor*: Anchors embryo to surrounding embryo sac and ovule tissue; serves as a conduit for nutrients from maternal sporophyte into developing proembryo. Suspensor senesces after the heart stage and is not a functional part of the mature seed.
- Embryonic organs and tissue-types differentiate during globular-heart transition phase.

Most of embryo is devoted to formation of the root. The hypophysis (Hs) gives rise to the root meristem.

*Hypophysis*: The uppermost cell of the suspensor from which part of the root and rootcap in the embryo are derived.

The shoot further differentiates into the axis and cotyledon. After the critical heart embryo stage, the shoot meristem is determined. The shoot meristem will give rise to the epicotyl, which in turn, will mature into the bulk of the plant body.

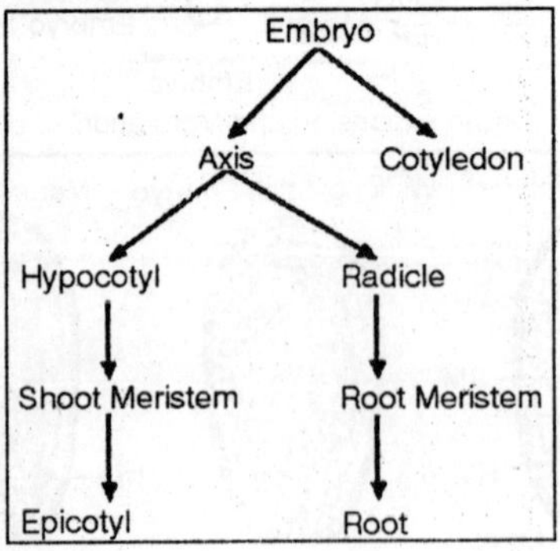

## CLONING AND CLONING VECTORS

A cloning vector is a small piece of DNA into which a foreign DNA fragment can be inserted. The insertion of the fragment into the cloning vector is carried out by treating the vehicle and the foreign DNA with the same restriction enzyme, then ligating the fragments together. There are many types of cloning vectors. Genetically engineered plasmids and bacteriophages (such as phage λ) are perhaps most commonly used for this purpose. Other types of cloning vectors include bacterial artificial chromosomes (BACs) and yeast artificial chromosomes (YACs).

Most commercial cloning vectors have key features that have made their use in molecular biology so widespread. In the case of expression vectors, the main purpose of these vehicles is the controlled expression of a particular gene inside a convenient host organism (eg. E. coli). Control of expression can be very important; it is usually desirable to insert the target DNA into a site that is under the control of a particular promoter. Some commonly used promoters are T7 promoters, lac promoters (bla promoter) and cauliflower mosaic virus's 35s promoter (for plant vectors).

To allow for convenient and favourable insertions, most cloning vectors have had nearly all their restriction sites engineered out of them and a synthetic multiple cloning site (MCS) inserted that contains many restriction sites. MCSs allow for insertions of DNA into the vector to be targeted and possibly directed in a chosen orientation. A selectable marker, such as an antibiotic resistance [eg. beta-lactamase (see figure)] is often carried by the vector to allow the selection of positively transformed cells (see Screening below). All plasmids must carry a functional origin of replication (ORI; not shown in figure).

Some other possible features present in cloning vectors are: vir genes for plant transformation, intergrase sites for chromosomal insertion, lacZα fragment for α complementation and blue-white selection, and/or reporter genes in frame with and flanking the MCS to facilitate the production of recombinant proteins [eg. fused to the Green fluorescent protein (GFP) or to the glutathione S-transferase.

### Properties of all Cloning Vectors

- Will accept foreign DNA and can still complete their life cycle
- Permit those bacterial or yeast cells that contain the foreign DNA to be distinguished from those that do not

### Types of Vectors

- Plasmids.

- Bacteriophage.
- Cosmids.
- Yeast artificial chromosomes (YACs).
- Bacterial artificial chromosomes (BACs).

### Insert Size of Various DNA Cloning Vectors

| | Vector | Insert size (kb) |
|---|---|---|
| | Plasmid | <10 kb |
| | Bacteriophage | 9-20 kb |
| | Cosmids | 33-47 kb |
| | BACs | 75-125 kb |
| YACS | 100-1000 kb | |

## PLASMIDS

### Characteristics of Plasmids

- Extrachromosomal circular DNA molecules which are not part of the bacterial genome.
- Size range: 1-200 kb.

3. carry functions advantageous to the host such as.
   - Produce enzymes which degrade antibiotics or heavy metals
   - Produce restriction and modifying enzymes

- Replication is coupled to host replication in a:
  - Stringent manner: One (or two) plasmids made during each round of bacterial replication
  - Relaxed manner: 10-200 copies of the plasmid made during each round of bacterial replication; (this can be increased to 1000-2000 plasmids by stopping host protein synthesis and replication with the antibiotic chloramphenicol)

Foreign DNA is inserted into a plasmid (or any cloning vector) by ligating the DNA into a complementary site in the plasmid. These sites are generated by digesting the DNA and vector with the same restriction enzyme. (The site for the restriction enzyme that is chosen should only be represented once in the plasmid. Thus, when the plasmid is digested, a single, linear molecule would be generated.) The foreign DNA is then inserted into the plasmid by the action of the enzyme DNA ligase. The next step is to insert the ligated DNA into a bacterial cell for propagation. This is done by a technique called transformation. Bacterial cells are treated with either Ca2Cl or Rb2Cl.

This treatment produces pores in the bacterial cell wall and membrane through which the plasmid enters. Although there is no size limitation to the ligation reaction, transformation efficiency is dictated by the size of the plasmid.

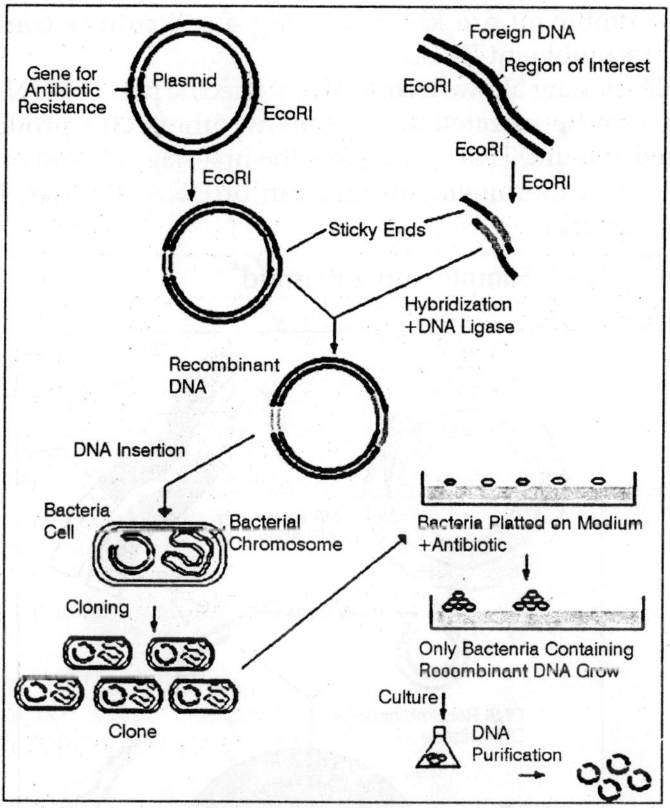


Fig. Cloning into a Plasmid

Process by which a plasmid is used to import recombinant DNA into a host cell for cloning. Many diseases are caused by gene alterations. Our understanding of genetic diseases was greatly increased by information gained from DNA cloning. In DNA cloning, a DNA fragment that contains a gene of interest is inserted into a cloning vector or plasmid.

The plasmid carrying genes for antibiotic resistance, and a DNA strand, which contains the gene of interest, are both cut with the same restriction endonuclease. The plasmid is opened up and the gene is freed from its parent DNA strand.

They have complementary "sticky ends." The opened plasmid and the freed gene are mixed with DNA ligase, which reforms the two pieces as recombinant DNA Plasmids + copies of the DNA fragment produce quantities of recombinant DNA. This recombinant DNA stew is allowed to transform a bacterial culture, which is then exposed to antibiotics.

All the cells except those which have been encoded by the plasmid DNA recombinant are killed, leaving a cell culture containing the desired recombinant DNA.

DNA cloning allows a copy of any specific part of a DNA (or RNA) sequence to be selected among many others and produced in an unlimited amount. This technique is the first stage of most of the genetic engineering experiments: production of DNA libraries, PCR, DNA sequencing, *et al.*

## Inserting a DNA Sample into a Plasmid

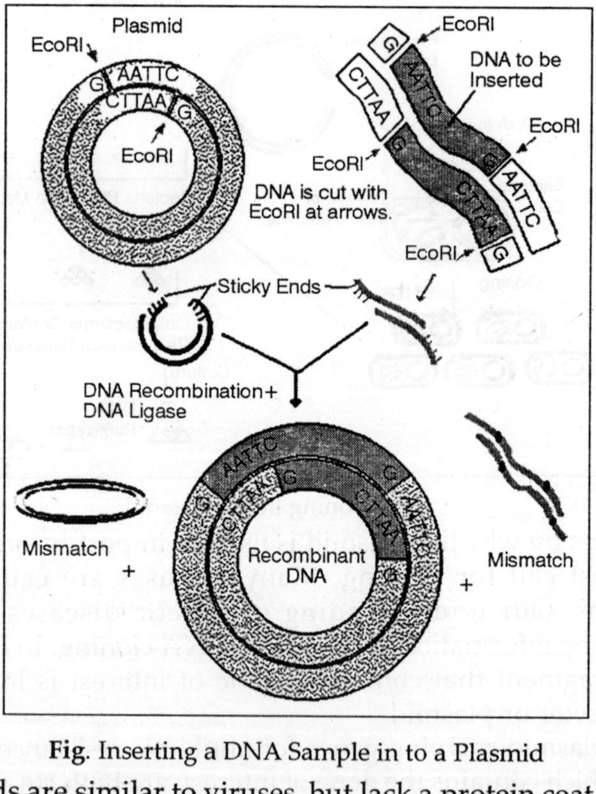


**Fig.** Inserting a DNA Sample in to a Plasmid

Plasmids are similar to viruses, but lack a protein coat and cannot move from cell to cell in the same fashion as a virus. Plasmid vectors are small circular molecules of double stranded DNA derived from natural plasmids that occur in bacterial cells. A piece of DNA can be inserted into a plasmid if both the circular plasmid and the source of DNA have recognition si:es for the same restriction endo:uclease.

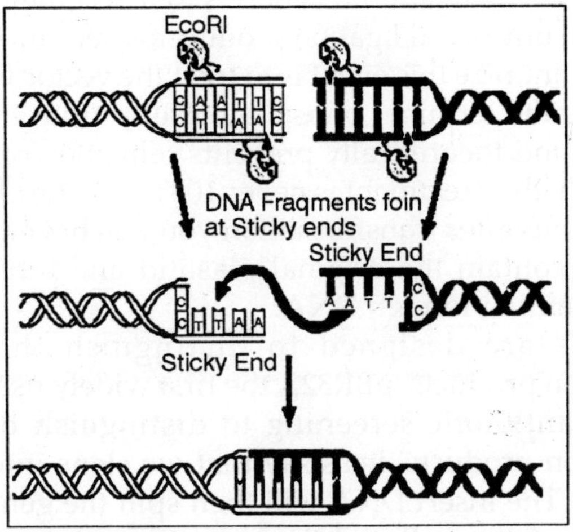


**Fig.** Restriction Enzyme Action of EcoRI

Restriction enzymes, also called restriction nucleases (EcoRI in this example), surrounds the DNA molecule at the point it seeks(sequence GAATTC). It cuts one strand of the DNA double helix at one point and the second strand at a different, complementary point (between the G and the A base). The separated pieces have single stranded "sticky-ends," which allow the complementary pieces to combine. The newly joined pieces are stabilized by DNA ligase. EcoRI, one of many restriction enzymes, is obtained from the bacteria Escherichia coli.

The plasmid and the foreign DNA are cut by this restriction endonuclease (EcoRI in this example) producing intermediates with sticky and complementary ends. Those two intermediates recombine by base-pairing and are linked by the action of DNA ligase. A new plasmid containing the foreign DNA as an insert is obtained. A few mismatches occur, producing an undesirable recombinant. The new plasmid can be introduced into bacterial cells that can produce many copies of the inserted DNA . This technique is called DNA cloning.

**Effect of Plasmid Size on Transformation Efficiency**

| Molecule size (kb) | % Maximum probability |
|---|---|
| 2.0 | 57 |
| 3.2 | 100 |
| 4.3 | 86 |
| 12.5 | 43 |
| 20.0 | 36 |
| 39.0 | 14 |
| 54.0 | 6 |

The goal of the ligation reaction is to insert the foreign DNA into the vector. An unwanted ligation product also occurs - religated vector. One way to minimize this event is to treat the vector with phosphatase. This removes the terminal phosphate group from the restriction site of the vector and theoretically prevents religation of the two ends. In reality though, this treatment is never 100%, so a low level of religation does occur. Thus after transformation you will have two types of cells: those which contain the original plasmid and those that contain a plasmid containing foreign DNA.

Plasmids are designed to distinguish the two types of transformation products. pBR322, the first widely used vector, utilizes differential antibiotic screening to distinguish the two types of transformation products. Let's say that we clone into the BamHI site of the vector. The insert DNA will then split the gene responsible for tetracycline resistance. But at the same time, the gene for ampicillin resistance is left intact. Transformed cells are first grown on bacterial plates containing ampicillin. This will kill all the cells that do not contain a plasmid. But we still cannot say which cells contain foreign DNA. Those cells that grew on ampicillin are then replica plated on plates with ampicillin and tetracycline. Those cells which grow in the presence of the ampicillin, but die under tetracycline selection contain plasmids which have foreign DNA inserts.

pBR322 was a breakthrough for molecular biology, but the double screening procedure was time consuming and could be subject to error. In 1981, a new series of plasmids were developed that permitted the identification of the foreign DNA containing cells in a single screening step. These are called the pUC plasmids. As with pBR322, ampicillin resistance is used as one selectable marker. The second marker is based on insertional inactivation of the E. coli lacZ gene. The wild type gene can hydrolyze a specific dye [X-Gal (5-bromo-4-indoyl-B-D-galactopyranoside)] to a blue colour, and the bacterial colony is stained blue. A multiple cloning site has been inserted into this gene. This site will accept fragments ending in a number of different restriction enzymes. Upon insertion of DNA into this site, the activity of the gene is eliminated and the colony appears white in colour. Thus transformed colonies containing plasmids with inserts can be distinguished from those with plasmids without inserts based on the colour of the colony and the ability of the colony to grow on an ampicillin containing media.

## cDNA Cloning (Cloning Eukaryotic mRNA)

cDNA cloning is a method of obtaining a DNA copy of the mRNAs that are expressed at a specific stage in the development of the plant.

In this manner, you can enrich the library that you will screen for those sequences in which you are interested. The reagent required for this type of cloning approach is mRNA. These mRNAs have a poly A+ tail. This tail permits the isolation of poly A+ mRNA by using either oligo-dT or oligo-U columns. Total RNA is run through one of these columns under conditions which favour the binding of the tail to the matrix on the column. After the column is extensively washed, the conditions are changed and the bound mRNA is isolated. This is the starting reagent for cDNA cloning.

## Steps in cDNA Cloning

- Bind oligo-dT to the poly +A tail of the mRNA.
- Add reverse transcriptase and make a DNA copy of mRNA (cDNA). Erase the mRNA with alkali and high temperature. (First strand synthesis)
- Add a C-tail to the 3' end of the cDNA with terminal transferase.
- Add oligo-dG to the tailed-cDNA and make the second strand with reverse transcriptase or the Klenow fragment of DNA Polymerase I.
- Add dC's to the 3' end of the double-stranded cDNA with terminal transferase
- Add C-tailed, ds-cDNA to G-tailed, PstI cut pBR322 and anneal. (DNA ligase is not required for this step.)
- Transform E. coli cells.
- Select Tetr/Amps cells. Most, but not all inserts can be removed by PstI digestions.

## BACTERIOPHAGE LAMBDA VECTORS

We have talked about plasmids as vectors for cloning small pieces of DNA. The limitation of this vector is the size of DNA that can be introduced into the cell by transformation. This presents problems when you are trying to create a genomic library of a large genome such as with plants. A genomic library contains all of the DNA found in the cell of the plant (or any organism). If you digest plant DNA to completion with a restriction enzyme, ligate those fragments into a plasmid vector and transform bacterial cells, only a portion of those fragments will be represented in the final transformation products. If a gene of interest is located on a large fragment then you will not be able to isolate that gene from a plasmid library. But what can be done to increase the probability of obtaining a clone which contains the entire gene. First you need to use a vector that can accept large fragments of

DNA. Examples of these are bacteriophage and cosmid vectors and more recently yeast artificial chromosomes. Bacteriophage lambda vectors were developed because several observations were made that suggested that they could complete their life cycles even if foreign DNA was inserted into a portion of its genome. This suggested that certain regions of the virus were not essential. Let's first discuss the life cycle of lambda.

- *Adsorption*: The phage particle binds at a maltose receptor site of the bacterial cell; growing the cell in the presence of the sugar increase the number of receptor sites
- *Penetration*: DNA is injected into the cell; at this point it can enter one of two pathways;
  - *Lysogenic pathway*: The phage DNA becomes integrated into the genome and is replicated along with the bacterial DNA; it remains integrated until it enters the lytic pathway
  - *Lytic pathway*: Large scale production of bacteriophage particles that eventually leads to the lysis of the cell; base pairing at the cos site leads toa circular molecule.

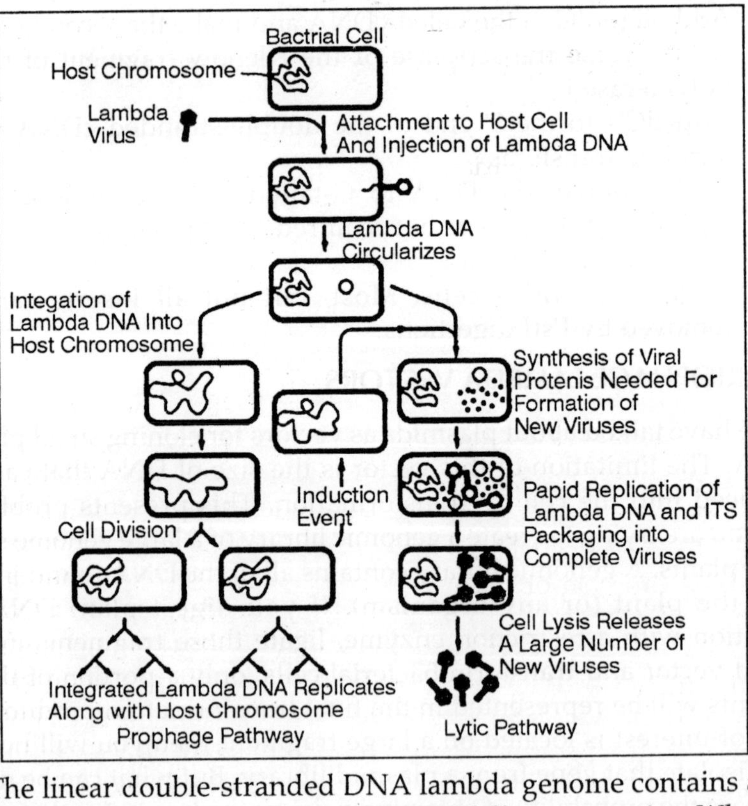


The linear double-stranded DNA lambda genome contains about 50,000 nucleotide pairs and encodes 50-60 different proteins. When the lambda DNA enters the cell the ends join to form a circular DNA molecule. The bacteriophage can multiply in E. coli by a lytic pathway, which destroys the cell, or it can enter a latent prophage state. Damage to a cell carrying a lambda prophage induces the prophage to exit from the host chromosome and shift to lytic growth (green arrows). The entrance and exit of the lambda DNA from the bacterial chromosome are site-specific recombination events

- *Early transcription*: Transcription proceeds from the pL and pR promoters, through the N and cro genes and stops at terminators tL and tR1; a low level of transcription through the O and P genes occurs and terminates at tR2; the N product is an antitermination factor that is important for the next stage of transcription.
- *Delayed early transcription*: The N product binds to RNA polymerase and transcription proceeds past the tL, tR1 and tR2 terminators; genes to the left of N, involved in recombination, to the right of cro, involved in replication, are expressed at this point; another protein expressed from the Q gene is used for antitermination of later transcription.
- *Replication*: Early replication is through a theta form initiated from a single origin of replication site; later replication is via rolling circle replication; this produces long concatamers of the phage DNA that are cleaved at the cosL and cosR sites.
- *Late transcription*: The protein product of the cro gene builds up to a critical level and then binds to the oL and oR to stop early transcription; another protein, a product of the Q gene, has built up and activates transcription at the p'R promoter by antitermination; transcription terminates with in the b region; this transcription results in the production of the proteins required for the head and tail of the mature phage particle and those required for bacterial cell lysis.
- *Assembly*: A prophage head is produced; a unit length DNA is placed into the head by the action of the Nu1 and A proteins; the DNA is locked into place by the D protein and ter function of the A protein clips the DNA at the cosL and cosR sites; the concatamer is released, the tail is added and the mature phage particle is completed.

Packaging of the DNA into the head does not require a complete length of wild type lambda. It has been determined that a lambda molecule that is between 78% and 105% of wild type length can be packaged. This is from 37 to 53 kb in length. Two important developments suggested that lambda may be suitable as a cloning vector.

First it was determined that the gene products between the J and N genes could be removed and the life cycle could be completed. Second, restriction enzyme sites could be eliminated which permitted the development of a vector with a single site for insertion of foreign DNA.

*Two types of vectors have been developed*:

- *Insertional vector*: DNA is inserted into a specific site.
- *Replacement vector:* Foreign DNA replaces a piece of DNA (stuffer fragment) of the vector.

Let's talk about a specific vector EMBL 3 and EMBL 4. One important concern when cloning with lambda vectors is that you want to maximize the number of resulting phage particles that contain foreign DNA. Or said another way you want to minimize the number of wild type particles. One approach is through spi selection. This refers to sensitivity to P2 interference.

**Table. Bacteriophage Phenotypes for Growth on P2 Lysogens**

| Phenotype | Growth on P2 Lysogens? (bacterial strain) |
|---|---|
| spi+ (red+gam+) | Poor |
| spi– (red–gam–) | Good |

EMBL 3/4 vectors have placed the red and gam genes in the stuffer fragment. Thus only those particles from which the stuffer has been replaced can grow well in a P2 lysogen bacterial cell.

## Cloning in Lambda Vectors

Phages are viruses that can infect bacteria. The major advantage of the phage vector is its high transformation efficiency, about 1000 times more efficient than the plasmid vector.

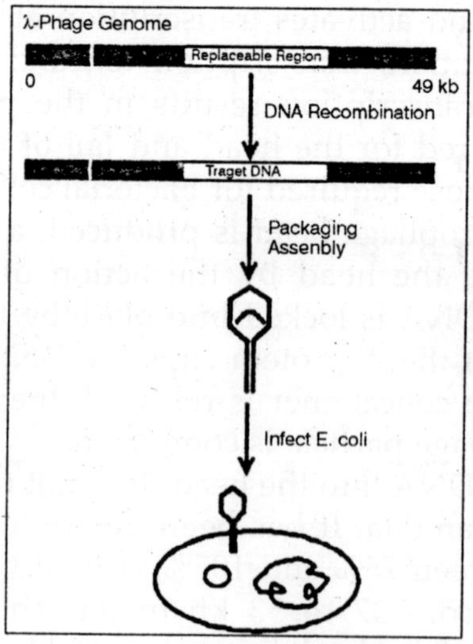


**Fig**. Schematic Drawing of the DNA Cloning Using ? Phages as Vectors.

The DNA to be cloned is first inserted into the DNA, replacing a nonessential region. Then, by an in vitro assembly system (described below), the virion carrying the recombinant DNA can be formed. The genome is 49 kb in length which can carry up to 25 kb foreign DNA.

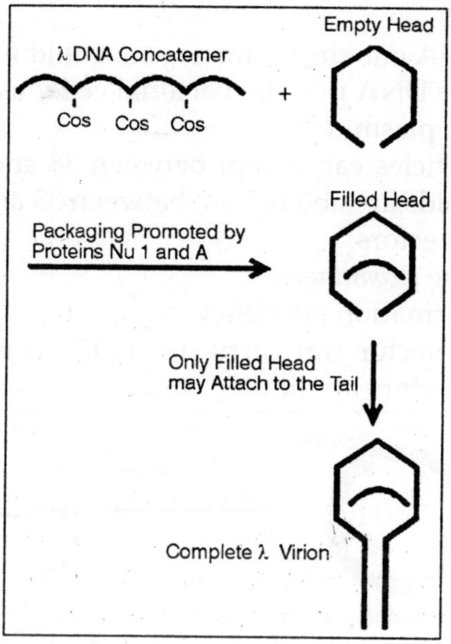


**Fig.** The Assembly Process of the Virion.

The extreme ends of the DNA are known as COS sites, each is single stranded, 12 nucleotides long. Because their sequences are complementary to each other, one end of DNA may base-pair with the other end of a different DNA, forming concatemers. The two ends of a DNA may also bind together, forming a circular DNA. In the host cell, the DNA circularizes because ligase may seal the join of the COS sites. In the assembly process of virions, two proteins Nu1 and A can recognize the COS site, directing the insertion of the DNA between them into an empty head.

The filled head is then attached to the tail, forming a complete virion. The whole process normally takes place in the host cell. However, to prepare the virion carrying recombinant DNA, the following in vitro assembly system is commonly used. Proteins Nu1 and A are encoded by the genes in the genome. If the two genes are mutated, DNA cannot be packaged into the pre-assembled head. Because tails attach only to filled heads, the cell will accumulate separate empty heads and tails, which can then be extracted. When the extract is mixed with recombinant DNA and proteins Nu1 and A, the complete virion carrying recombinant DNA will be assembled.

## COSMIDS

Cosmids are plasmid vectors that contain cos sites. The cos site is the only requirement for DNA to be packaged into a phage particle. Cosmids were developed in light of this observation. How do you clone into cosmid vectors?

- Clone the DNA into the vector as you would with any plasmid.
- Introduce the DNA into the bacterial cell via a phage particle.
- Propagate as plasmid.

Since phage particles can accept between 38 and 53 kb of DNA and since most cosmids are about 5 kb, between 33 and 48 kb of DNA can cloned in these vectors.

*It has the following advantages*:

- High transformation efficiency.
- The cosmid vector can carry up to 45 kb whereas plasmid and phage vectors are limited to 25 kb.

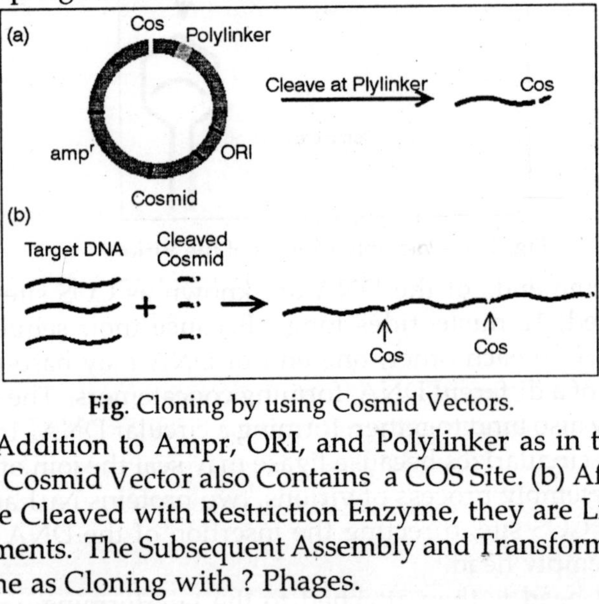


**Fig.** Cloning by using Cosmid Vectors.

(a) In Addition to Ampr, ORI, and Polylinker as in the Plasmid Vector, the Cosmid Vector also Contains a COS Site. (b) After Cosmid Vectors Are Cleaved with Restriction Enzyme, they are Ligated with DNA Fragments. The Subsequent Assembly and Transformation Steps are the same as Cloning with ? Phages.

### Problems Associated with Lambda and Cosmid Cloning

- Since repeats occur in eukaryotic DNA rearrangements can occur via recombination of the repeats present on the DNA inserted into lambda or cosmid.
- Cosmids are difficult to maintain in a bacterial cell because they are somewhat unstable.

## YEAST ARTIFICIAL CHROMOSOMES (YACS)

One goal of molecular genetics is to obtain physical data about the genomic organization of long stretches of DNA. Traditionally, this data has been obtained by a technique called chromosome walking. Walking is performed by subcloning the end of a lamda or cosmid clone and screening a library for other clones that contain similar sequence information. If this new clone overlaps a portion of the original clone, then the length of the DNA of interest is extended by the length of DNA in the second clone that is not found in the original clone.

By performing these steps successive times, a long distance map can be obtained. This technique though has difficulties. First, each step is technically slow. Second, if you use lambda or cosmid clones, you might only extend the region of interest by 5-10 kb in each step of the walk. Finally, if any of the clones that are obtained contain repeated sequences, the subclone could lead you to another region of the genome that is not contiguous with the region of interest.

Yeast artificial chromosomes can alleviate some of these problems because of the large (100-1000kb) amount of DNA that can be cloned. First of all, YACs cannot speed up each step of the walk because the subcloning and screening steps can only be performed so quickly. But they can solve the other two problems. Because they carry large amounts of DNA, each step can easily extend the region of interest by 50-100 kb and up to as much as 500 kb. Thus a long distance map of the region can be obtained in several steps. Secondly, although repetitive regions may be 10-20 kb in length they are rarely, longer than 50 kb. Thus a YAC with 100kb will contain some region that is single copy which can be used for further steps in the walk.

*Features of YACs*:

- Large DNA (>100 kb) is ligated between two arms. Each arm ends with a yeasttelomere so that the product can be stabilized in the yeast cell. Interestingly, larger YACs are more stable than shorter ones, which favours cloning of large stretches of DNA.
- One arm contains an autonomous replication sequence (ARS), a centromere (CEN) and aselectable marker (trp1). The other arm contains a second selectable marker (ura3).
- Insertion of DNA into the cloning site inactivates a mutant expressed in the vector DNA and red yeast colonies appear.
- Transformants are identified as those red colonies which grow in a yeast cell that is mutant for trp1 and ura3. This ensures that the cell has received an artificial chromosome with both telomeres (because of complementation of the two mutants) and the artificial chromosome contains insert DNA (because the cell is red).

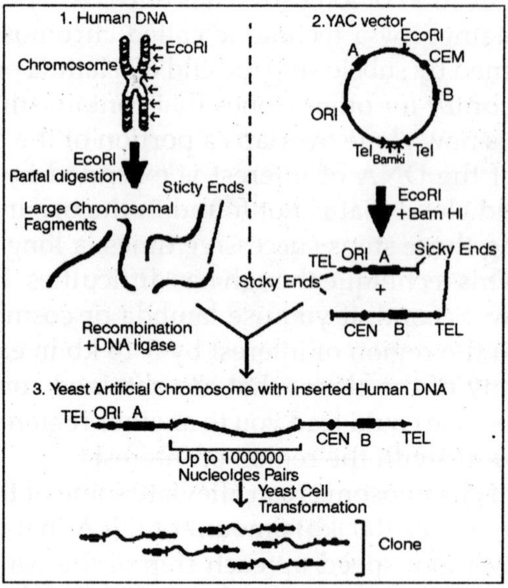


Fig. Cloning into a Yeast Artificial Chromosome (YAC)

DNA cloning is a technique that allows the wholesale production of a specific DNA sequence. DNA containing a gene of interest is inserted into the purified DNA genome of a self replicating element, which can be a plasmid, a virus or in this case, a yeast artificial chromosome (YAC). A YAC can be considered as a functional artificial chromosome (self replicating element), since it includes three specific DNA sequences that enable it to propagate from one cell to its offspring:

- *TEL*: The telomere which is located at each chromosome end, protects the linear DNA from degradation by nucleases.
- *CEN*: The centromere which is the attachment site for mitotic spindle fibers, "pulls" one copy of each duplicated chromosome into each new daughter cell.
- *ORI*: Replication origin sequences which are specific DNA sequences that allow the DNA replication machinery to assemble on the DNA and move at the replication forks.

*It also contains few other specific sequences like*:

- *A and B*: selectable markers that allow the easy isolation of yeast cells that have taken up the artificial chromosome.
- Recognition site for the two restriction enzymes EcoRI and BamHI.

While DNA cloning into a plasmid allows the insertion of DNA fragment of about 10,000 nucleotide base pairs, DNA cloning into a YAC allows the insertion of DNA fragments up to 1,000,000 nucleotide base pairs. Why is it so important to be able to clone such large sequences? To map the entire human genome (3x1,000,000,000 nucleotide base pairs) it would require more than 100,000 plasmid clones. In principle, the human genome could be represented in about 10,000 YAC clones.

## BACTERIAL ARTIFICIAL CHROMOSOMES (BACS)

A bacterial artificial chromosome (BAC) is a DNA construct, based on a functional fertility plasmid (or F-plasmid), used for transforming and cloning in bacteria, usually E. coli. F-plasmids play a crucial role because they contain partition genes that promote the even distribution of plasmids after bacterial cell division. The bacterial artificial chromosome's usual insert size is 150-350 kbp, but can be greater than 700 kbp. A similar cloning vector called a PAC has also been produced from the bacterial P1-plasmid.

Bacterial Artificial Chromosomes (BAC) have been developed to hold much larger pieces of DNA than a plasmid can. BAC vectors were originally created from part of an unusual plasmid present in some bacteria called the F' plasmid. The F' plasmid allows bacteria to have "sex" (well, sort of: F' helps bacteria give its genome to another bacteria but this only happens rarely when bacteria are under a lot of stress). F' had been studied extensively and it was found that it could hold up to a million basepairs of DNA from another bacteria. Also, F' has origins of replication and bacteria have a way to control how F' is copied. In 1992, Hiroaki Shizuya took the parts of F' that were important, cleaned it up, and turned it into a vector.

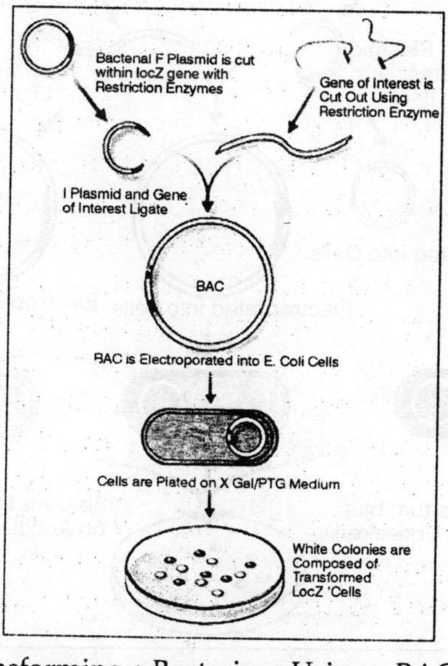


Fig. Transforming a Bacterium Using a BAC Vector

BAC vectors are able to hold up to 350 kb of DNA and have all of the tools that a vector needs to work properly, like replication origins, antibiotic resistance genes, and convenient places where clone DNA can insert itself. With these vectors it is possible to study larger genes, several genes at once, or entire viral genomes. By using a vector that can hold larger pieces of DNA, the number of clones required to cover the human genome six times theoretically could drop from 1.8 billion to about 50 million.

Researchers have modified BAC vectors to become more convenient to use and more useful in specialized situations. In addition to the antibiotic resistance gene that was added to identify transfected bacteria, a gene was added that enabled the bacteria to turn the colourless substance X-gal/IPTG blue. This substance is found in the chicken-soup-like media on which the bacteria colonizes.

This colour-changing gene, called lacZ, is split apart when the clone DNA is incorporated into the vector, so it is possible to tell not only if a bacteria had been transfected (meaning incorporated into the cell), but also if the bacteria was transfected with the vector containing insert DNA or just the vector alone (remembering that if the vector has properly incorporated the clone DNA, it will have lost its ability to change X-gal/IPTG).

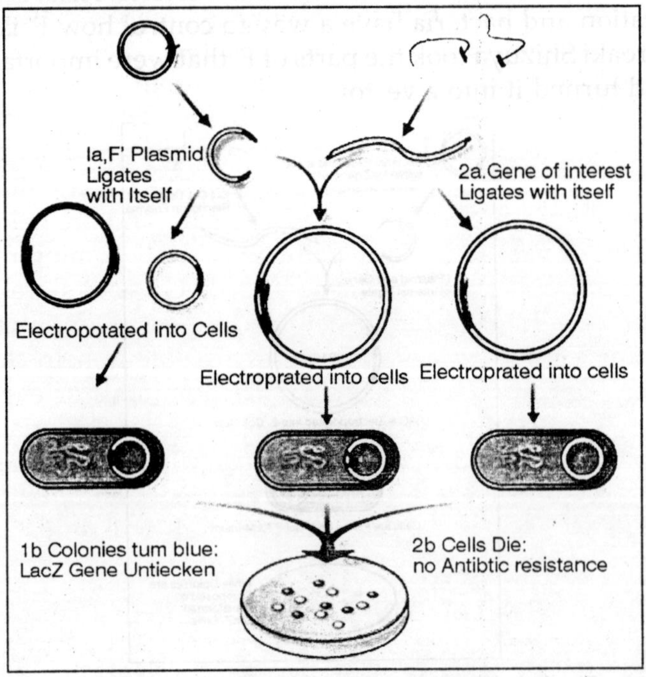


**Fig.** Selecting for Transformed Bacteria

Another similar modification uses the gene called sacB, which encodes a protein called levansucrase. This protein turns sucrose - table sugar - into levan, a toxic substance to bacteria. In a similar fashion, bacteria grown on media with sucrose will die if sacB is not broken up by inserted DNA. If the vector carries a DNA insert, then sacB is broken and it won't produce levansucrase, so the bacteria can survive in the sucrose media. In theory, only bacteria transfected with a vector containing insert DNA would be able to grow and form colonies.

Some modifications to BAC vectors make them more specialized. For instance, there are researchers studying the herpes virus who have made a BAC vector that can be cultured in bacterial cells and then when put into a mammalian cell, instantly releases its insert DNA - in this case, the whole herpes virus genome. With such a vector, it is easier to grow sufficient amounts of the herpes virus for research, since it can live in bacterial cultures, instead of requiring their endemic mammalian cell cultures, which are extremely difficult to maintain. This in turn makes it easier to make modifications to the virus and study what each of its genes do. The creation of BAC vectors has allowed researchers to do many things that they could not do before and do them more quickly and more easily. Some of the scientific practices that have simplified with BAC vectors includes phylogenetic studies (studies which examine species' relationship to one another) and what the absolute minimum size of a genome could be.

The world is currently overrun by a plethora of microbial species that cannot be grown in cultures. BACs have allowed researchers to look at microbial DNA without having to actually grow the organisms, since the DNA is kept within easy-to-grow bacterial cultures. BAC vectors are also useful for studying pathogens, and are helpful in the development of vaccines. Many pathogens are becoming resistant to all antibiotics available to medicine and BAC vectors are playing a tremendous role in discovering new and powerful antibiotics in the environment. Discovering enzymes that are able to help clean up oil spills or help farmers breed healthier farm animals or even process radioactive waste are just a few examples of what Bacterial Artificial Chromosomes can do. BAC vectors are useful tools and the methods for working with them are fairly well developed. Continuing advances such the modification of experimental animals will only increase the wide variety of uses for big bad BACs.

## LIBRARY SCREENING AND GENE SEQUENCING

Once a library is constructed it is screened for a particular gene of interest. Screening is based on homology between a probe and one of the clones in the library. The probe is normally a nucleic acid that has some sequence homology to the gene that is represented in the library. The library is the collection of clones from the source DNA which is inserted into a cloning vector. For example, a bean genomic lambda library contains pieces of the entire complement of bean DNA, from 9-20 kb in size, cloned into a lambda vector. Any probe used to screen the library should have some homology to a clone in the bean library, and this homology would allow you to select the appropriate clone from the many clones that do not contain your DNA of interest.

Probes come in several forms. The more homologous (similar in DNA sequence) the probe is to the sequence that is being sought, the easier it is to select a clone from the library. For example, a bean lambda library is considered a genomic library because it contains all the DNA sequences found in the bean genome. A bean leaf cDNA library, though, would contain only those sequences that are expressed in the leaf after they have undergone processing.

Thus the cDNA clones would not contain any of the intron sequences or the controlling elements of the gene. To obtain a genomic clone from the bean lambda library, the best probe would be a cDNA clone obtained from screening a bean cDNA library. This type of probe, a probe that contains the exact sequence of the sequence that is being sought, is called a homologous probe. But how was the original cDNA clone obtained so that it could be used as a probe.

The cDNA library from which the clone was obtained might have been screened by a probe from another species which represents coding information for the same gene but from another species. For example, a bean leaf cDNA library may be screened with a tomato RUBISCO small subunit clone. A probe that contains DNA which encodes for the gene of interest but from another species is called a heterologous probe. Many plant genes have been cloned by screening libraries with both homologous and heterologous probes.

Polymerase chain reaction techniques are now commonly used to clone genes. To use this technique primers must be designed that are complementary to your target sequence. One oligonucleotide will be complementary to the anticoding strand (the DNA strand of the gene complementary to the mRNA and used as a template for transcription). The second oligonucleotide will be complementary to the coding strand (the DNA strand of the gene complementary to the anticoding strand).

If the gene has been cloned in another species, you can use that sequence information to design primers to amplifiy a fragment of the gene from the DNA of your species. Often, the gene you have an interest in has never been cloned, but you may have isolated the protein.

If this is the case, the first step of this method is to obtain partial protein sequence information of the protein. Micropeptide sequencers are available that can rapidly generate sequence information. From this sequence, you can use reverse translation of the amino acid sequence to obtain the nucleic acid sequences of an amino-terminal fragment and a carboxy-terminal fragment. In this case the fragment complementary to anticoding strand will be a direct conversion of the genetic code for the amino acids in the amino-terminal fragment of the protein. The strand complementary to the coding strand will be complementary to the derived nucleic acid sequence of the carboxy-terminal fragment. Let's say that the following is the N-terminal sequence of the peptide for which you want to derive a synthetic oligonucleotide:

$NH_3$+–Met–Cys–Val–Lys–Thr–Pro–$CO_2$–

The following would be an appropriate probe for the gene.

5'-A T G T G T/G G T N A A A/G A G N C C – 3'

(N = A, T, G, and C)

When you are constructing these probes several concepts should be kept in mind. First, the genetic code is written in the mRNA sequence. This synthetic oligonucleotide will thus, be complementary to the anit-coding strand that is used as the template for the mRNA. Secondly, because you only know the amino acid and not the DNA sequence you must deal with redundancies of the genetic code.

For example leucine can be represented by CTA, CTT, CTG, CTC. Thus your synthetic oligonucleotide needs to contain all the possibilities. Therefore the above sequence is actually a mixture of 64 (4×4×2×2) different sequences. Because of this redundancy it is always best to make you oligonucleotide one short of full length.

This will help some of the redundancy problems. In this example, the probe is 17 nucleotides long and the redundancies for proline are eliminated. Another method of reducing the redundancy problem is to incorporate the nucleotide deoxyinosine at any position where a high level or redundancy occurs. Deoxyinosine is a modified nucleotide that does not pair with any of the four bases found in DNA. Therefore, it does not affect the homology of the probe.

*The following may be the amino acid sequence of a carboxy-terminal fragment*:

$NH_3$+–Phe–Tyr–His–Thr–Val–Ala–$CO_2$–

The appropriate oligonucleotide sequence from the above sequence for PCR amplification would be:

5'– G C N A C N G T A/G T G A/G T A A/G A A -3'

Amplification of DNA between these two oligonucleotides will provide a homologous probe because it will represent the exact sequence of the gene for which you are searching. Once this probe is radiolabelled by nick-translation (as all probes must be radilabelled before library screening), it can be used to screen a cDNA or genomic library for a clone complementary to the amplified fragment.

## DNA Sequencing With the Chain-Termination, Dideoxy Technique of Sanger

Once you have a candidate clone, you will want to sequence it to determine if it is similar to another gene already sequenced or whether it is unique. DNA sequencing is a DNA replication based reaction. The two requirements for DNA replication are a DNA template and a free 3'-OH group. These requirements must be met for any sequencing procedure. The following steps illustrate the Sanger procedure, the most widely used DNA sequencing procedure.

- Although this technique originally used DNA cloned in a vector that can generate single- stranded DNA, sequencing is now routinely performed in double-stranded plasmid or cosmid vectors. pUC plasmids and related plasmids are popular for double stranded sequencing. To begin the procedure, the DNA is made single stranded by a combination of heating and alkaline conditions.
- A primer which has a free 3'-OH group is annealed to the vector. A convenient primer is the multiple cloning site from the pUC plasmids. DNA synthesis will initiate here. 3. Four reactions are performed, each containing dATP, dGTP, TTP and 32P-dCTP and one dideoxy nucleotide in low concentration. Dideoxy nucleotides have a hydrogen rather than a - OH group attached to the #3 carbon and cannot be used for further extension of the chain. Thus in this reaction, various synthesis products will end with that nucleotide.
- Add DNA polymerase to the reaction. Nucleotides will be added one at a time but the chain will terminate when a dideoxy nucleotide is inserted. Because the dideoxy nucleotide is limiting a series of products, each ending with the same base, will be obtained.
- Separate the fragments on an acrylamide gel, develop via autoradiography and read the sequence.

## Chapter 13

# Map-based or Positional Cloning

The first step of map-based or positional cloning is to identify a molecular marker that lies close to you gene of interest. This procedure typically is done my first finding a marker in the vicinity of the gene (several cM away). For the initial screening smaller population sizes are used (60-150 individuals) The next step is to saturate the region around that original molecular marker with other markers. At this point you are looking for a one that rarely shows recombination with your gene. At this stage, the population size could increase to 300-600 individuals.

The next step is to screen a large insert genomic library (BAC or YAC) with your marker to isolate clones that hybridize to your molecular marker. Once you identify the initial markers that map are near (or better yet) flank your gene and fournd a a clone to which the markers hybridize, you are on your way to determining where that gene resides. The steps that follow are termed chromosomal walking. This procedure involves creating new markers (usually sequences at the end of the clone) and screening your segregating population with these new markers.

Often this population is large (1000-3000 individuals). The goal is to find a set of markers that co-segegate (no recombination) with your gene of interest. Co-segregation means that whenever one allele of your gene is expressed, the markers associated with that allele are also present. In other words, recombination is not seen between your gene and the markers. If these markers do not cosegregate, you select new large insert clones and repeat the process until you have a clone whose markers co-segregates with your gene. To speed the cloning process, it is best to begin with a marker that is tightly linked to the gene with which your are working. Therefore you will not have to do a lot of additional screening.

Because you have your gene flanked on a single clone between two markers, you now know that the gene must be between those two markers. DNA fragments between the flanking markers are cloned and introduced into a genotype mutant for your gene by a genetic engineering technique called plant transformation. If transgenic plant expresses the wild type phenotype, you then know the gene of interest is on that fragment. At this point you must sequence the fragment to find a potential open reading frame (ORF), sequences that most likely will encode a gene product.

In the best situation, only a single ORF is found, but this often is not the case. Usually several possible ORFs are found and new transgenic plants are created by transforming with a single ORF. Once this ORF is shown to rescue the mutant phenotype, you then perform an in-depth molecular and biochemical analysis of newly cloned gene.

*These steps can be summarized as follows*:

- Identify a marker tightly linked to your gene in a "large" mapping population.
- Find a YAC or BAC clone to which the marker probe hybridizes.
- Create new markers from the large-insert clone and determine if they co-segregate with your gene.
- If necessary, re-screen the large-insert genomic library for other clones and search for co-segregating markers.
- Identify a candidate gene from large-inset clone whose markers co-segregate with the gene.
- Perform genetic complementation (transformation) to rescue the wild-type phenotype.
- Sequence the gene and determine if the function is known.

If you do not have a transformation system for your species of interest, the confirmation is much more difficult. The most powerful approach in this case is recombinational or mutant analysis. Lets say you have two mutants in your gene. You could cross these two mutants and search a large segregating population for an individual with a restored normal phenotype. Next you would clone from each of these three individuals (the two mutants and the restored line) homologouse sequences that you feel contains the correct gene. If you can show that the restored line contains the same sequence as your predicted gene, and that the two mutants have unique changes in the gene sequence not found in the normal gene, you have obtained compelling evidence that the putative sequence is indeed your gene of interest.

## AN EXAMPLE OF PLANT MAP-BASED CLONING

Recently map-based cloning was employed to clone the first plant resistance gene that follows the gene-for-gene interaction. The tomato gene that was cloned was Pto and it provides resistance against bacterial speck disease of tomato caused by Pseudomonas syringae pv. tomato. This pathogen expresses the avirulence gene avrPto that interacts with Pto to provide resistance in tomato. This gene was cloned by Martin *et al.* [Science (1993) 262:1432].

*The following is the list of the various steps Martin et al. used to clone the Pto gene.*

- A genetic population of 251 F2 plants was screened with DNA probes, and the locus TG538 cosegregated with Pto.
- A YAC library was screened with the TG538 probe, and the clone PTY538-1 was identified.
- Primers that marked the end of the 400 kb clone were created, and the population was screened. The left end of the PTY538-1 (PCR marker PTY538-1L was 1.8 cM from Pto, and the right end of the clone (PTY538-1R) cosegregated with Pto. The right end of the clone maybe to left or right of Pto, so it is important to determine if the YAC clone spans the Pto gene.
- 1300 plants from F2, F3 and cultivars were screened, and one plant that contains the Pto allele at the TG538 and the PTY538-1R pto allele was found. For this result to occur, Pto must be located on the YAC clone PTY538-1.
- To find a candidate clone on the YAC clone, DNA from the PTY538-1 clone was hybridized to a leaf cDNA library. 200 clones were found and 30 analysed. A population of 50 plants with recombination in the Pto region was screened. cDNA clone CD127 cosegregated with Pto.
- Because Southern hybridization with CD127 detected many polymorphic fragments, the clone could be detecting exons or a multigene family. 14 other cross hybridizing clones were found that all represented six different classes of related genes. Analyses of PTY538-1 detected all (but one) of the genomic fragments that were detected with Southern hybridization. It was concluded that PTY538-1 contained a multigene family.
- To determine if the cDNAs contained the Pto gene, transformation vectors were created with CD127 (1.2 kb insert; vector pPTC5) and CD186 (2.4 kb insert; vector pPTC8). Two pPTC8 resistant plants were found. No pPTC5 resistant plants were found.
- One resistant plant (genotype=Pto/Pto) was crossed to a susceptible cultivar (genotype= pto/pto). Of 22 plants 9 contained the CD186 sequence and were resistant to the bacterial speck pathogen, and the 13 without CD186 were susceptible to bacterial speck.

- The clone CD186 was sequenced, and a 321 amino acid opening reading frame was found. The sequence of the open reading frame was similar to a serine-threonine protein kinase.

The steps outlined for this experiment demonstrate the general approach needed to clone a gene using map-based cloning. The specific steps may vary between this and other map-cloning experiments, but the same general steps must be met. For example, finding a cosegregating clone may require a high density map developed with RAPD or AFLP markers. The availability of techniques to screen a high volume of potential markers has changed one aspect of map-based cloning. It was originally thought that it would be necessary to perform a chromosome walk into the gene. The major difficulty with this technique in plants is the abundance of repetitive sequence. If the walk went through such a region, the walk could be diverted to a different chromosome, and the gene would not be found. This new technique of using high volume marker screening to move close to the gene has been called chromosome landing.

# Index